Springer Theses

Recognizing Outstanding Ph.D. Research

Aims and Scope

The series "Springer Theses" brings together a selection of the very best Ph.D. theses from around the world and across the physical sciences. Nominated and endorsed by two recognized specialists, each published volume has been selected for its scientific excellence and the high impact of its contents for the pertinent field of research. For greater accessibility to non-specialists, the published versions include an extended introduction, as well as a foreword by the student's supervisor explaining the special relevance of the work for the field. As a whole, the series will provide a valuable resource both for newcomers to the research fields described, and for other scientists seeking detailed background information on special questions. Finally, it provides an accredited documentation of the valuable contributions made by today's younger generation of scientists.

Theses are accepted into the series by invited nomination only and must fulfill all of the following criteria

- They must be written in good English.
- The topic should fall within the confines of Chemistry, Physics, Earth Sciences, Engineering and related interdisciplinary fields such as Materials, Nanoscience, Chemical Engineering, Complex Systems and Biophysics.
- The work reported in the thesis must represent a significant scientific advance.
- If the thesis includes previously published material, permission to reproduce this must be gained from the respective copyright holder.
- They must have been examined and passed during the 12 months prior to nomination.
- Each thesis should include a foreword by the supervisor outlining the significance of its content.
- The theses should have a clearly defined structure including an introduction accessible to scientists not expert in that particular field.

More information about this series at http://www.springer.com/series/8790

Ali Kemal Yetisen

Holographic Sensors

Doctoral Thesis accepted by
the University of Cambridge, UK

 Springer

Author
Dr. Ali Kemal Yetisen
Department of Chemical Engineering
and Biotechnology
University of Cambridge
Cambridge
UK

Supervisor
Prof. Christopher R. Lowe
Department of Chemical Engineering
and Biotechnology
University of Cambridge
Cambridge
UK

ISSN 2190-5053 ISSN 2190-5061 (electronic)
Springer Theses
ISBN 978-3-319-13583-0 ISBN 978-3-319-13584-7 (eBook)
DOI 10.1007/978-3-319-13584-7

Library of Congress Control Number: 2014956221

Springer Cham Heidelberg New York Dordrecht London

Printed on acid-free paper

Springer International Publishing AG Switzerland is part of Springer Science+Business Media (www.springer.com)

Parts of this thesis have been published in the following journal articles:

1. Yetisen AK, Naydenova I, Vasconcellos FC, Blyth J, Lowe CR (2014) Holographic Sensors: Three-Dimensional Analyte-Sensitive Nanostructures and their Applications. Chem Rev 114 (20):10654–10696
2. Yetisen AK, Montelongo Y, da Cruz Vasconcellos F, Martinez-Hurtado JL, Neupane S, Butt H, Qasim MM, Blyth J, Burling K, Carmody JB, Evans M, Wilkinson TD, Kubota LT, Monteiro MJ, Lowe CR (2014) Reusable, robust, and accurate laser-generated photonic nanosensor. Nano Lett 14 (6):3587–3593
3. Yetisen AK, Butt H, da Cruz Vasconcellos F, Montelongo Y, Davidson CAB, Blyth J, Chan L, Carmody JB, Vignolini S, Steiner U, Baumberg JJ, Wilkinson TD, Lowe CR (2014) Light-Directed Writing of Chemically Tunable Narrow-Band Holographic Sensors. Adv Opt Mater 2 (3):250–254
4. Yetisen AK, Qasim MM, Nosheen S, Wilkinson TD, Lowe CR (2014) Pulsed laser writing of holographic nanosensors. J Mater Chem C 2 (18):3569–3576
5. Yetisen AK, Martinez-Hurtado JL, Garcia-Melendrez A, Vasconcellos FC, Lowe CR (2014) A smartphone algorithm with inter-phone repeatability for the analysis of colorimetric tests. Sens Actuators, B 196 (0):156–160
6. Tsangarides CP, Yetisen AK, da Cruz Vasconcellos F, Montelongo Y, Qasim MM, Wilkinson TD, Lowe CR, Butt H (2014) Computational modelling and characterisation of nanoparticle-based tuneable photonic crystal sensors. RSC Adv 4 (21):10454–10461
7. Yetisen AK, Akram MS, Lowe CR (2013) Paper-based microfluidic point-of-care diagnostic devices. Lab Chip 13 (12):2210–2251

Supervisor's Foreword

As the world's population surpasses 7 billion, healthcare systems around the word face unique challenges. North America, Western Europe and Japan have ageing populations, which are a growing concern due to the increasing demand for long-term care limited by the shortages in healthcare workers. On the contrary, the developing world is populated with younger inhabitants; however, 55 % of the inhabitants of the developing world live in rural regions, where infrastructure is scarce and healthcare equipment is outdated. Unsurprisingly, the developing world countries face healthcare challenges to protect their population from infectious and non-communicable diseases. Hence, these global healthcare trends require efficient medical services and technologies that can meet the unfulfilled demand of ever-growing populations.

At the heart of the healthcare systems is screening large populations to monitor high-risk individuals and develop epidemiological strategies to timely mitigate emerging epidemics. When diseases are diagnosed at an early stage, the treatment is often simpler and more likely to be effective. Hence, the innovation in rapid, accurate diagnostic devices with connectivity has the potential to reduce the burden on the healthcare systems and patients worldwide. In their development, point-of-care diagnostic devices play a unique role since they are lightweight, portable and can be made readily available to healthcare workers and patients. Monitoring conditions and diseases rapidly at point-of-care offers unique opportunities in personalised medicine, which may allow optimisation of therapies, and subsequently produce improved treatment options. Such portable diagnostics can also allow efficient management of chronic diseases, where frequent measurements and treatments are required. In the developing world, low-cost diagnostics can reach underserved regions and reduce the poverty-related diseases to empower communities.

The development of point-of-care diagnostic devices concerns both the study of sensors and readout devices, and their clinical evaluation and the social context of use. These devices need to be user-friendly, fool-proof, lightweight, have a long shelf life and offer connectivity with emerging mobile devices. However, commercial sensors are complicated by the power-consuming electronics and custom readout devices, which increase the cost per diagnosis. Many colorimetric tests result in erroneous results due to subjective interpretation and have limitations in colorimetric range, in which the colour code differs from one assay to another. The development of an easy-to-interpret, colorimetric and quantitative sensing platform can standardise the readouts for visual interpretation, and facilitate simultaneous detection of conditions and diseases while also offering the possibility to quantify the assay by smartphones and wearable devices.

Ali Yetisen is a polymath whose research spans both physical and social sciences including point-of-care diagnostics, micro/nanofabrication, optical devices, microfluidics, smartphone apps, commercialisation, entrepreneurship, patent law and FDA regulations. His doctoral thesis makes a contribution to the development of reusable colorimetric optical sensors for applications in point-of-care diagnostics. His thesis harnesses laser-light writing in functionalised hydrogels for the production of holograms that allow quantification of analytes in aqueous solutions. Holographic sensor development approaches outlined include silver-halide chemistry, laser ablation and photopolymerisation. The fabricated sensors allow quantification of pH, organic solvents, metal ions and glucose. The present work is supplemented with computational simulations to lay a foundation for the laser writing techniques and principle of operation of the sensors to enhance our knowledge of how hydrogel-based sensing materials function. For example, the thesis describes the development of a kinetic theory for the hydrogel-based sensors in order to reduce the readout time. The work also shows a clinical trial to test the performance of the holographic sensors for the analysis of glucose in the urine samples of diabetic patients. In the development of readout technologies, the thesis demonstrates a smartphone application for the quantitative analysis of various colorimetric pH, protein and glucose assays. The final chapter of the thesis critically reviews the efforts in holographic sensor development, points out the limitations and draws guidelines for the future work.

This thesis not only shows a viable strategy for the fabrication and optimisation of holographic sensors in the entire visible spectrum, but it also demonstrates new insights into their functioning. The findings of this thesis provide a sound basis for the development of optimised holographic sensors as a step towards producing multiplexed diagnostic devices that can meet user requirements at point-of-care.

Cambridge, UK, November 2014   Prof. Christopher R. Lowe

Abstract

Developing non-invasive and accurate diagnostics that are easily manufactured, robust and reusable will provide monitoring of high-risk individuals in any clinical or point-of-care environment, particularly in the developing world. There is currently no rapid, low-cost and generic sensor fabrication technique capable of producing narrow-band, uniform, reversible colorimetric readouts with a high-tunability range. This thesis presents a theoretical and experimental basis for the rapid fabrication, optimisation and testing of holographic sensors for the quantification of pH, organic solvents, metal cations and glucose. The sensing mechanism was computationally modelled to optimise its optical characteristics and predict the readouts. A single pulse of a laser (6 ns, 532 nm, 350 mJ) in holographic "Denisyuk" reflection mode allowed rapid production of sensors through silver-halide chemistry, in situ particle size reduction and photopolymerisation. The fabricated sensors consisted of off-axis Bragg diffraction gratings of ordered silver nanoparticles and localised refractive index changes in poly(2-hydroxyethyl methacrylate) and poly-acrylamide films. The sensors exhibited reversible Bragg peak shifts, and diffracted the spectrum of narrow-band light over the wavelength range $\lambda_{peak} \approx 500–1,100$ nm. The application of the holographic sensors was demonstrated by sensing pH in artificial urine over the physiological range (4.5–9.0), with a sensitivity of 48 nm/pH unit between pH 5.0 and 6.0. For sensing metal cations, a porphyrin derivative was synthesised to act as the crosslinker, the light absorbing material, the component of a diffraction grating as well as the cation chelating agent. The sensor allowed reversible quantification of Cu^{2+} and Fe^{2+} ions (50 mM–1 M) with a response time within 50 s. Clinical trials of a glucose sensor in the urine samples of diabetic patients demonstrated that the glucose sensor has an improved performance compared to a commercial high-throughput urinalysis device. The experimental sensitivity of the glucose sensor exhibited a limit of detection of 90 µM, and permitted diagnosis of glucosuria up to 350 mM. The sensor response was achieved within 5 min and the sensor could be reused about 400 times without compromising its accuracy. Holographic sensors were also tested in flake form, and integrated with paper-iron oxide composites, dyed filter and chromatography papers, and nitrocellulose-based test strips. Finally, a generic smartphone application was developed and tested to quantify colorimetric tests for both Android and iOS operating systems. The developed sensing platform and the smartphone application have implications for the development of low-cost, reusable and equipment-free point-of-care diagnostic devices.

Contents

1 Point-of-Care Diagnostics 1
    1.1 The Development of Rapid Diagnostics 3
    1.2 Sensing Mechanisms 7
        1.2.1 Colorimetric Reagents 7
        1.2.2 Electrochemical Sensors. 12
        1.2.3 Colloidal Nanoparticles (NPs). 13
        1.2.4 Chemiluminescence (CL) 13
        1.2.5 Electrochemiluminescence (ECL) 14
        1.2.6 Fluorescence 15
        1.2.7 Genetically-Engineered Cells 15
    1.3 Next Generation Diagnostics 15
    References. 17

2 Fundamentals of Holographic Sensing 27
    2.1 Fabrication of Optical Devices 27
    2.2 History of Holography. 28
    2.3 The Origins and Working Principles of Holographic
        Sensors 32
    2.4 Computational Modelling of Holographic Sensors
        in Fabrication and Readout 37
        2.4.1 Photochemical Patterning. 37
        2.4.2 Simulations of the Optical Readouts 39
    2.5 Conclusions 45
    References. 46

3 Holographic pH Sensors 53
    3.1 Holographic pH Sensors via Silver-Halide Chemistry 53
    3.2 Fabrication of Holographic pH Sensors Through
        in Situ Size Reduction of Ag^0 NPs 55

3.3 Characterisation of Holographic pH Sensors.............. 56
    3.3.1 Microscopic Imaging of Holographic pH Sensors 57
    3.3.2 Effective Index of Refraction Measurements........ 62
    3.3.3 Angular-Resolved Measurements 63
    3.3.4 Diffraction Efficiency Measurements............. 64
    3.3.5 Polymer Thickness and Roughness Measurements. ... 65
3.4 Optical Readouts 66
    3.4.1 Holographic pH Sensors Fabricated Through
        Silver Halide Chemistry...................... 66
    3.4.2 Holographic pH Sensors Fabricated Through
        in Situ Size Reduction of Ag^0 NPs 68
    3.4.3 Interference Due to Metal Ions 69
    3.4.4 Ionic Strength Interference in pH Measurements..... 70
    3.4.5 Sensing pH in Artificial Urine 71
    3.4.6 Paper-Based Holographic pH Sensors 72
3.5 Discussion..................................... 77
References.. 80

4 Holographic Metal Ion Sensors 85
4.1 Fabrication of Holographic Metal Ion Sensors
    via Photopolymerisation........................... 86
4.2 Optical Readouts 89
    4.2.1 Organic Solvents in Water.................... 89
    4.2.2 Quantification of Cu^{2+} and Fe^{2+} Ions
        in Aqueous Solutions 89
4.3 Conclusions.................................... 93
References.. 94

5 Holographic Glucose Sensors. 101
5.1 Diabetes Mellitus 101
5.2 Holographic Glucose Sensors........................ 103
5.3 Computational Modelling of Holographic Glucose Sensors. ... 106
5.4 Fabrication of Holographic Glucose Sensors.............. 107
5.5 Holographic Glucose Sensors for Urinalysis.............. 108
    5.5.1 Holographic Glucose Sensor Readouts............ 110
    5.5.2 Holographic Glucose Sensor Readouts
        in Artificial Urine........................ 112
    5.5.3 Lactate and Fructose Interference 115
    5.5.4 Interference Due to Osmolality................. 117
    5.5.5 Tuning of the Wavelength Shift Range
        of the Holographic Glucose Sensor.............. 118
    5.5.6 Exposure Bath to Tune the Base Position
        of the Bragg Peak......................... 118

5.6 Kinetic Theory for Hydrogel Swelling................... 120
5.7 Quantification of Glucose Concentration in Urine.......... 122
5.8 Lactate and Fructose Interference 125
5.9 Conclusions.................................. 128
References... 130

6 Mobile Medical Applications........................... 135
6.1 Global Health and Mobile Medical Applications........... 135
6.2 A Smartphone Algorithm for the Quantification
of Colorimetric Assays 138
6.2.1 Calibration of the Application.................. 138
6.2.2 User Interface of the Smartphone Application........ 140
6.2.3 Colorimetric Measurements 141
6.3 Conclusions.................................. 145
References... 146

7 The Prospects for Holographic Sensors 149
7.1 The Development of Fabrication Approaches 149
7.2 Ligand Chemistry 152
7.3 Multiplexing Holographic Sensors with Microfluidic
Devices..................................... 155
7.4 Readouts with Smartphones and Wearable Devices......... 156
7.5 The Vision for Holographic Sensors 158
References... 159

Abbreviations

Å	Angstrom
AAm	Acrylamide
Ag^+	Silver ion
Ag^0	Silver metal
$AgNO_3$	Silver nitrate
$AgClO_4$	Silver perchlorate
AgBr	Silver bromide
Au^0	Gold metal
ATP	Adenosine-5'-triphosphate
ADP	Adenosine diphosphate
A^-	Conjugate base of the acid
ATMA	(3-Acrylamidopropyl)trimethylammonium chloride
3-APB	3-(Acrylamido)phenylboronic acid
2-APB	2-Acrylamidophenylboronate
5-F-2-MAPBA	2-Acrylamido-5-fluorophenylboronic acid
a.u.	Arbitrary units
$\sim$	Approximately
°C	Celsius
CL	Chemiluminescence
CCD	Charge-coupled device
CH_4N_2O	Urea
CI_{low}	Lower confidence bound
CI_{high}	Upper confidence bound
CH_3CN	Acetonitrile
CMOS	Complementary metal-oxide-semiconductor
CIE	International Commission on Illumination
CVD	Chemical vapour deposition
ø	Diameter
Λ or d	Lattice spacing between the two consecutive layers
d_{ks} and d_{kss}	Shortest distances to the sample point

D	Dimension
DBAE	2-(Dibutylamino)-ethanol
$\Delta\lambda$	Changes in Bragg peak position
Δn_0	Changes in effective refractive index
$\Delta\Lambda$	Changes in grating period
$\Delta\theta$	Changes in the Bragg angle
DMPA	2,2-Dimethoxy-2-phenylacetophenone
DCC	N,N'-dicyclohexylcarbodiimide
DMAP	4-(Dimethylamino)pyridine
DMSO	Dimethyl sulphoxide
DCM	Dichloromethane
DI	Deionised
DNA	Deoxyribonucleic acid
$\eta(t)$	Number of molecules bound
Ery	Erythrocyte
Eq	Equation
ECL	Electrochemiluminescence
EDMA	Ethylene dimethacrylate
ESEM	Environmental scanning electron microscopy
FT-IR	Fourier transform infrared spectroscopy
FWHM	Full width at half maximum
g	Gram
GOx	Glucose oxidase
G-6-P	Glucose-6-phosphate
G-6-PDH	Glucose-6-phosphate dehydrogenase
h	Hour
hv	Incident light
H^+	Hydrogen ion
H_z	Magnetic field strength
H_{oz}	Initial magnetic field strength
HCl	Hydrochloric acid
HeNe	Helium Neon
HEMA	2-Hydroxyethyl methacrylate
^{1}H NMR	Proton Nuclear Magnetic Resonance
HK	Hexokinase
HA	Protonated form of the acid
HL7	Health level seven
I	Normalised intensity distribution
I_{max}	Maximum intensity
IMCI	Integrated Management of Childhood Illness
iOS	Internet operating system
IDA	Iminodiacetic acid
k	Propagation constant or Integer (see context)
kV	Kilovolt
KOH	Potassium hydroxide

L	Litre or Free macrocyclic ligand (see context)
λ_{peak}	Bragg peak of the 1st order diffracted light in vacuo
λ_h	Change in the periodicity of the multilayer structure
λ_∞	Wavelength at the infinite
λ_{shift}	Step Bragg peak shift
$\Delta\lambda$	Bragg peak shift
λ_0	Initial wavelength
LASER	Light Amplification by Stimulated Emission of Radiation
LED	Light-emitting diode
LiBr	Lithium bromide
LDH	Lactate dehydrogenase
Lue	Leucocyte
LVDT	Linear variable differential transformer
m	Metre
M	Molar
min	Minute
mJ	Millijoule
MBAAm	N,N'-methylenebisacrylamide
MAA	Methacrylic acid
MHz	Megahertz
M^+	Solvated metal ion
M^+-L	Metal-macrocyclic ligand pair
M^+L	Contact pair
$(ML)^+$	Final complex
MP	Megapixel
n	Effective index of refraction of the recording medium
ns	Nanosecond
n	Sample size
N	Newton or Number of functional groups (see context)
$n(t)$	Rate of change of bound molecules
N_g	Total number of glucose molecules
N_f	Total number of boronic acid groups
NP	Nanoparticle
Na_2HPO_4	Sodium phosphate dibasic
$(NH_4)_2SO_4$	Ammonium sulphate
Nd:YAG	Nd-Yttrium-Aluminum-Garnet
$Na_2S_2O_3$	Sodium thiosulphate
Na_2CO_3	Sodium carbonate
Na_2HPO_4	Sodium phosphate dibasic
NaOH	Sodium hydroxide
NAD^+	Nicotinamide adenine dinucleotide
NADH	Nicotinamide adenine dinucleotide hydride
$NaHCO_3$	Sodium bicarbonate
OD	Optical density
pK_a	Acid disassociation constant

Ph	Photochemistry
PBS	Phosphate buffered saline
PMMA	Poly(methyl methacrylate)
PVA	Poly(vinyl alcohol)
PBG	Photonic band gap
PDMS	Poly(dimethylsiloxane)
pHEMA	Poly(2-hydroxyethyl methacrylate)
pAAm	Poly(acrylamide)
R	Ratio of the intensities of the reference and the object
r	Position
RGB	Red, green, blue
R_c, G_c, B_c	Non-linear red, green and blue values
R_l, G_l, B_l	Linearised the red, green and blue values
R^2	Correlation coefficient
RCA	Rolling circle amplification
RI	Refractive index
rpm	Revolution per minute
s	Second
σ	Standard deviation (see context)
s_y	Standard of residuals
s_m	Standard of slope
s_b	Standard of intercept
Si	Silicon
SDK	Software development kit
SEM	Scanning electron microscopy
t	Thickness
θ	Angle of incidence of illumination or tilt angle (see context)
THF	Tetrahydrofuran
TEM	Transmission electron microscopy
TACPP	5,10,15,20-Tetrakis[4″-(3‴-(acryloyloxy) propoxy)phenyl-4′-carboxyphenyl] porphyrin
U	Uncertainties
UV	Ultra Violet
v	Volume
$v_\pm$	±Variation
4-VPBA	4-Vinylphenylboronic acid
W	Watt
WD	Working distance
WHO	World Health Organisation
x_m, y_m	2D chromaticity values
X, Y, Z	Tristimulus values
y	Constant position

Chapter 1
Point-of-Care Diagnostics

Rapid tests that are low-cost and portable are the first line of defence in healthcare systems. Dipstick and lateral-flow are the two universal assay formats as they are lightweight and compact, and provide qualitative results without external instrumentation. However, existing formats have limitations in the quantification of analyte concentrations. Hence, the demand for sample preparation, improved sensitivity and user-interface has challenged the commercial products. Recently, capabilities, sensors and readout devices were expanded to multiplexable assays platforms, which might transcend the capabilities of existing design format of diagnostic tests. This chapter outlines the evolution of diagnostic devices and current trends in the development of qualitative and quantitative sensing devices for applications in healthcare, veterinary medicine, environmental monitoring and food safety. The chapter also discusses design parameters for diagnostics, their functionalisation to increase the capabilities and the performance, emerging sensing platforms and readout technologies. The factors which limit the emerging rapid diagnostics to become commercial products are also discussed.

The life expectancy has grown worldwide, which also increased the healthcare spending [1, 2]. For example, 720 million people will be aged 65 or older by 2020. Currently four in five people over the age of 75 take at least one prescribed medicine, and this trend is set to increase [3]. For some of the medications, large pharma find it the hard to recover R&D costs while the pressure from the regulatory agencies also increased for the use of new drugs as the first line of defence based on efficacy and cost. Hence, the global healthcare trends and ever increasing pressure from regulatory agencies put a strong case for the development of diagnostics for healthcare monitoring as well as the evaluation of drug efficacies in clinical trials. All these considerations parallels governments' and insurance companies' interests in obtaining the best performance possible and treatment benefits they support. These trends are behind the driving force for the development of diagnostics that can reduce the healthcare costs by developing effective drugs and identifying diseases and conditions at an early stage.

The major stumbling block in monitoring and controlling diseases/contaminations remains delivering simple, low-cost and robust diagnostic tests [4, 5]. In the developing world, the basic healthcare infrastructure and trained healthcare personnel are

A.K. Yetisen, *Holographic Sensors*, Springer Theses,
DOI 10.1007/978-3-319-13584-7_1

limited [6, 7]. Other trends include increase in healthcare associated infections [8, 9], preservation of the life span of cost-effective drugs [10], increase in spurious/counterfeit medicines [11], and mitigating the epidemic-to-pandemic transitions of infectious diseases [12]. In terms of impact, low-cost diagnostics will reach underserved communities. Low-cost diagnostics that can allow local communities in developing regions to improve healthcare [13], environmental safety [14], animal health [15, 16], and food quality [17] will play key roles in the United Nation's Post-2015 Millennium Development Goals.

In the developing world, medical diagnostics for poverty-related conditions are outdated. In the absence of the diagnostic equipment, healthcare personnel make their decisions based on symptoms. WHO's Integrated Management of Childhood Illness (IMCI) is a diagnosis guideline based on signs and symptoms with minimum or no instrumentation [18, 19]. However, such systematic guidelines have limitations in (i) distinguishing asymptomatic diseases, (ii) detecting multiple infections, (iii) identifying the disease window period, and (iv) quantifying concentrations of target analytes. Eventually, the development and delivery of affordable testing technologies can enable local communities that lack access to technical and human resources present in urban areas. The deficiency in diagnostics and healthcare resources can have an irreversible negative effect on developing economies. Such an economic impact is pronounced for non-infectious as well as infectious diseases. For example, the Ebola virus epidemic in West Africa has overwhelmed the healthcare systems of Liberia, Sierra Leone, Guinea and their neighbouring countries since 2013. Such outbreaks have a profound impact on the development of emerging economies. Currently, it is difficult to identify Ebola because its symptoms such as fever are generic, also seen in commonly occurring diseases such as malaria and typhoid fever. Existing laboratory techniques are based on transcription polymerase chain reaction (RT-PCR) and quantitative PCR; however, they are not portable and affordable in the developing world. Other rapid tests suffer from sensitivity and selectivity, which may put the entire healthcare system at risk by misdiagnosing patients. Yet, recent Ebola epidemic is an example of many ongoing difficulties in healthcare systems and highlights the dire need for the development of low-cost rapid diagnostics.

While the main application of rapid tests is in medical diagnostics, such assays are also necessary for veterinary testing, environmental monitoring and food quality testing. Low-cost diagnostics would allow farmers and entrepreneurs in the developing world to assess the quality of their products and mitigate potential risks due to enterohemorrhagic strain of *E. coli* (O157:H7) and *Salmonella* in undercooked meat and poultry products [17]. Low-cost diagnostics are also required in testing water supplies. Inadequate environmental monitoring capabilities were highlighted by the cholera epidemic in Haiti after the earthquake in 2010 [20–24]. Although the cholera epidemic took most people by surprise, it wasn't totally unexpected since Haiti ranked last out of 147 countries surveyed in the 2002 Water Poverty Index [25]. Due to globalisation, such epidemics also concern the developed nations. The development of low-cost, rapid quantitative diagnostics will aid in screening large regions and populations. For the development of rapid diagnostics, WHO has outlined a set of criteria corresponding to the acronym ASSURED: Affordable, Sensitive, Specific, User-friendly, Rapid and Robust, Equipment free and Deliverable to those who need it [10]. For the developing world, the trend towards low-cost is a priority; however, clinically useful sensitivity and specificity from rapid diagnostics assays are also required. Operating with low-volume samples without manual manipulation, being portable and functioning without sharps such as needles are considered desirable characteristics of rapid assays. Hence, the development of robust diagnostic tests can enable individuals and local communities to monitor their health condition and timely mitigate the spread of diseases.

1.1 The Development of Rapid Diagnostics

Historical assay formats for point-of-care testing are dipsticks and lateral flow devices. In the 1950s, the earliest paper-based strip test emerged for the quantification of the concentration of glucose in urine, and this product was marketed in the 1960s [26]. The principle of operation of the assay was based on glucose oxidase, and developed colours were compared to a reference chart for interpretation. Today, commercial urinalysis strip tests are adapted for a wide range of analytes. Traditionally, these tests are semi-quantitatively read with a colorimetric chart or with detection equipment such as CLINITEK Status® + Analyzer (Siemens) or Urisys 1100® Urine Analyzer (Roche). In the 1950s, parallel to the development of strip tests, latex agglutination assays and radioimmunoassays were also developed [27, 28]. Since the 1970s, nitrocellulose matrices have been used as a substrate for molecular detection [29–31]. In the 1980s, serological lateral-flow tests emerged [32]. The most notable example is the human pregnancy test, which was derived from the development of hCG beta-subunit radioimmunoassay [33]. Since then, commercial rapid lateral flow assays expanded beyond clinical diagnostics to veterinary, food, environmental applications, bio-defence and drug abuse (Table 1.1). Rapid diagnostic tests often have lower specificity and sensitivity than their laboratory bench counterparts. The majority of these tests are simple and provide yes/no answer, where response time is critical to the user.

The commercial rapid tests have various geometries and configurations with/out housing units (Fig. 1.1). Figure 1.1d shows a multiplexed lateral-flow assay, which allows the analyses of a single sample simultaneously. Lateral-flow immunoassays have two major configurations: Direct (i.e. double antibody sandwich assays) and competitive (i.e. inhibitive) formats. The assay format typically consists of a number of segments: sample pad, conjugate pad, reaction membrane and an absorbent pad (Fig. 1.1e). These segments are supported by a backing card and enclosed in a plastic cassette (housing). Another format in rapid diagnostics is flow-through (vertical), which is relatively more complex than lateral flow format, and its execution requires (i) sample placement, (ii) washing and (iii) addition of analyte-colloidal gold conjugates [68] (Fig. 1.1f). Most of the commercial assays require sample preparation.

Table 1.1 Rapid diagnostic tests in the market

Test[a]	Application
Urinalysis	Metabolic disorders: Human chorionic gonadotropin (pregnancy) [34], (pH, glucose, protein, ketone, leukocytes, nitrite, blood, urobilinogen, bilirubin, specific gravity) [35–39], albumin to creatinine ratio [40, 41], and ascorbic acid [37, 39]
	Drug abuse[b]: Alcohol, amphetamines, barbiturates, benzodiaze-pines, buprenorphine, cocaine, ketamine, methamphetamines, methadone, morphine/opiates, oxycodone, phencyclidine, propoxyphene, 9-tetrahydrocannabinol (marijuana), and tricyclic antidepressants [42–45]
Immunoassays	Infectious diseases: *C. difficile*, *Cytomegalovirus*, dengue fever, *E. coli*, enterics, epstein barr virus, mononucleosis, giardiasis, herpes simplex virus, HIV, Lyme disease, malaria, measles, *S. aureus*, methicillin-resistant *Staphylococcus aureus* (MRSA), mumps, rubella, syphilis, toxoplasmosis, tuberculosis, varicella zoster, West Nile virus, hepatitis B/C, Chagas disease, *chlamydia*, cholera, hantavirus, leishmaniasis, leptospirosis, *Listeria*, and *H. pylori* [46–50]
	Respiratory diseases: Influenza (flu), Legionnaire's disease, pneu-monia, respiratory syncytial virus, streptococcal pharyngitis, and *Streptococcus pneumoniae* [46–48, 50]
	Cardiovascular condition: acute kidney injury, acute coronary syndrome, cyslipidemia, heart failure, oral anticoagulation, shortness of breath, and venous thrombosis [46]
	Oncology: Bladder, colon cancer, and colorectal cancer [46, 48]
	Women's health: Osteoporosis, ovulation, and preeclampsia [46]
Veterinary diagnostics	Canine: Blood, urobilinogen, bilirubin, protein, nitrite, ketones, glucose, pH, density, leukocytes, heartworm (*Dirofilaria immitis*), parvovirus, *Giardia*, Lyme disease, distemper virus, *coronavirus*, *Ehrlichia*, *Leishmania*, adenovirus, rotavirus, pancreatic lipase, relaxin, blood group typing, *Borrelia*, *Brucella*, c-reactive protein, *Leptospira*, progesterone, rabies, rheumatoid factor, and vaccina-tion status [51–54]
	Feline: Blood, urobilinogen, bilirubin, protein, nitrite, ketones, glucose, pH, density, leukocytes, immunodeficiency virus, leuke-mia virus, heartworm (*Dirofilaria immitis*), *Ehrlichia*, *Leishmania*, *Giardia*, parvovirus, infectious peritonitis, *Toxoplasma gondii*, relaxin, blood group typing, *Borrelia*, chlamydia, coronavirus, panleukopenia, progesterone, and vaccination status [51, 53, 54]
	Bovine: Alpha toxin, *Brucella*, *Chlamydophila*, *Clostridium perfringens*, coronavirus, *Cryptosporidium*, rotavirus, *E. coli* K99, *Crypto-sporidium parvum*, epsilon toxin, foot-and-mouth disease virus, IgG, *Leptospira*, *Mycobacterium bovis*, *Neospora*, progesterone, rabies, parainfluenza-3, and rotavirus [53]
	Swine: Aujeszky's disease, *Clostridium perfringens*, *Cryptospori-dium*, epidemic diarrhoea virus, rotavirus, alpha toxin, foot-and-mouth disease virus, progesterone, and transmissible gastroenteritis virus [53]

(continued)

Table 1.1 (continued)

Test[a]	Application
	Equine: *Borrelia*, rotavirus, *Clostridium perfringens*, IgG, pregnant mare serum gonadotropin, progesterone, tetanus, and troponin [53]
	Avian: Influenza, *Chlamydophila*, infectious bursal disease, and Newcastle disease [53]
	Small ruminants: Foot-and-mouth disease virus [53]
Food and beverage safety tests	Mycotoxins: Aflatoxins, deoxynivalenol (vomitoxin), fumonisins, zearalenone, ochratoxin, T-2 and HT-2 toxin, patulin, ergot alkaloids) [55–58]
	Food pathogens: *E.Coli* O157:H7, *Listeria*, *Staphylococcus aureus*, and *Salmonella Enteritidis* [55–58]
	Food allergens: Almond, brazil nut, casein, cashew/pistachio, coconut, hazelnut, lupin, macadamia nut, mustard, peanut, sesame, soy, walnut, whole egg, β-lactoglobulin, total milk, crustacea, and gliadin/gluten [55, 56]
	Genetically modified organisms: Bulk grain, seed and leaf, toasted meal, and corn Comb 7 Traits [55]
	Veterinary drug residues: Chloramphenicol, nitrofuran AMOZ/AOZ/AHD, clenbuterol, ractopamine, beta agonists, dexamethasone, ciprofloxacin, quinolones, β-Lactam antibiotics, flunixin, aminoglycoside, amphenicol, enrofloxacin, macrolide, sulphonamide, tetracycline, and melamine [55–57]
	Beverage: Methanol contamination, acetic acid, citric acid, D-glucose, D-fructose, lactose, milk (lactic acid, urea), and wine (acetic acid, total acidity, glucose, fructose, L-lactic acid, L-malic acid, and sucrose) [56]
	Pesticide residues: Organophosphates, thiophosphates, and carbamates [56]
	Seafood analysis: Amnesic shellfish poisoning (ASP), marine biotoxins (okadaic acid), histamine, and sulphite residues [56]
	Species identification: Pork, horse, beef, fish, goat, poultry, rabbit and sheep [56]
Environmental monitoring devices	Water testing: Algae, alkalinity, aluminium, ammonia, arsenic, bleach, boron, bromine, cadmium, calcium hardness, carbon dioxide, chelant, chloride, chlorine, chromate, chromium, conductivity, copper, cyanuric acid, cyanide, detergents, dissolved oxygen, faecal streptococci, *E.Coli* and faecal coliforms, filming amine, fluoride, formaldehyde, glutaraldehyde, hardness, hydrazine, hydrogen peroxide, iodine, iron, lead, magnesium, manganese, molybdate/molybdenum, morpholine, nickel, nitrate, nitrite, oil in water, organophosphate, dissolved oxygen, ozone, quaternary ammonium compounds, peracetic acid, pH, phosphate, phosphonate, polyphosphates, polyquat, potassium, *Pseudomonas aeruginosa*, salinity, silica, sulphate, sulphide, sulphite, turbidity, total dissolved solids, tannin/lignin, tolcide PS biocide, and zinc [59–64]
	Soil: Humus, organic matter, pH, and plant tissue macronutrient (texture) [60]

(continued)

Table 1.1 (continued)

Test[a]	Application
Biothreat detection	Anthrax, plague, tularaemia, ricin, botulinum toxin, *Staphylococcal enterotoxin B*, orthopox, *Brucella*, abrin, biological warfare simulants, nerve (G&V series), Category A-C biothreat agents, and blister (HD) agents, acids, bases, aldehydes and oxidisers [65–67]

[a] Require minimal sample preparation
[b] Urine collection and detection are often integrated in one cup

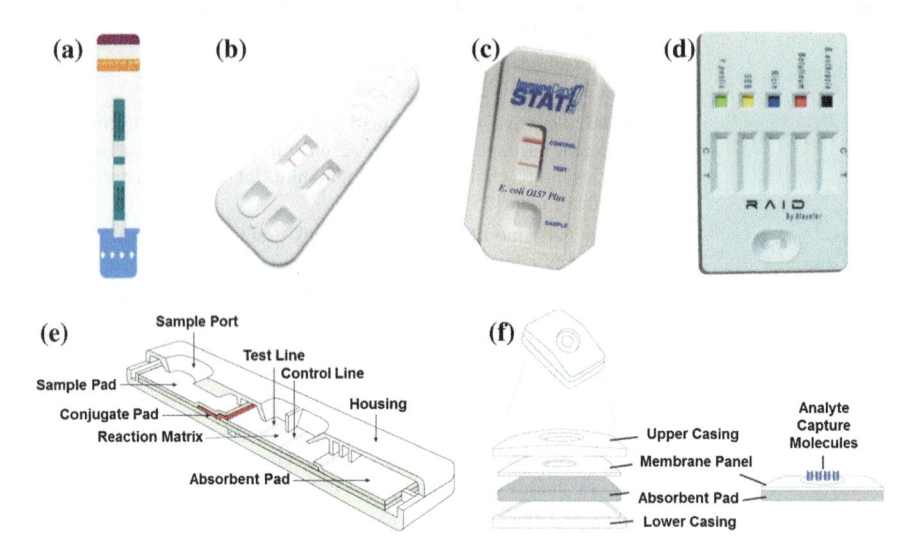

Fig. 1.1 Lateral-flow point-of-care assays in the market and their cassette formats. **a** Determine[TM] TB LAM Ag test.[©] (Alere), **b** Directigen[TM] EZ Flu A + B (Beckton Dickinson), **c** ImmunoCard STAT!® E. coli O157 Plus (Meridian Bioscience), **d** A multiplex lateral-flow assay. RAID[TM] 5 for biological threat detection (Alexeter Technologies), **e** Schematic of the lateral-flow assay, and **f** flow-through assay. Adapted from Ref. [69] with permission from The Royal Society of Chemistry

Typical approaches include the removal of contaminants from the samples to improve selectivity and sensitivity, and decrease the turnaround time. Additionally, the sample might require further processing to improve the signal-to-noise ratio. Sample preparation suffers from inhomogeneity, interfering agents, inhibitors, and it requires increasing the viscosity of samples such as whole blood and food samples. Increasing the sensitivity requires tedious sample preparation steps for low concentrations of target molecules or cells. The ideal sample preparation step should be cost-effective and is potentially executed in a single step. The desired outcome of the sample preparation is concentrating the target analyte(s) and reducing the background noise due to matrix interferences. Lateral-flow tests are low-cost, lightweight, portable, but the growing demand for higher sensitivity is challenging its current format [68]. For example, cardiac markers, biothreat and single cell detection require sensitivities that are not achievable with the existing assays. For these applications, lateral-flow tests must evolve and incorporate novel materials and labelling approaches. Progress in detection/quantification technologies, readout devices and manufacturing techniques will increase the reproducibility and sensitivity of lateral-flow assays. However, all these new advances must maintain the ultimate attractiveness of rapid tests: Sample-to-answer in a single step.

1.2 Sensing Mechanisms

A wide array of sensing mechanisms has been proposed for point-of-care diagnostic devices. Applications for such sensors include medical diagnostics, veterinary testing, environmental monitoring and food pathogen testing (Table 1.2). Sensing mechanisms demonstrated, with the exception of molecular dyes, required handheld readers. Excluding other types of sensors can be attributed to high costs, powering requirements and incompatibility with point-of-care assays.

1.2.1 Colorimetric Reagents

Molecular and enzymatic dyes are the simplest and most commonly used detection methods, which are semi-quantitatively interpreted by a colour reference chart. Urinalysis test strips such as Multistix (Siemens) and Chemstrip (Roche) are based on colorimetric reagents (Table 1.3). In advanced assays such as paper-based microfluidic devices, multiple detection zones are employed to capture different analytes within the same assay. Such multiplexed assays can be fabricated by printing wax channel-shaped patterns on paper/nitrocellulose, followed by heat treatment to form hydrophobic barriers in the matrix. The patterned regions can be spotted with enzymes, acid-base indicators or dyes in confined zones. Usually, a single inlet wicks the sample, which is distributed to the confined regions to react with the immobilised reagents. pH-, glucose- and protein- sensitive reagents have been used in paper-based microfluidic devices [70, 73]. In the glucose assay, a positive result is observed when the colour shifts from clear to brown due to the enzymatic oxidation of iodide to iodine. Similarly, a positive result in protein assays is interpreted from a colour change of tetrabromophenol blue from yellow to blue [70, 73]. Up to date, multiplex colorimetric assays included medical diagnostics [81, 87, 90, 91], environmental tests [76, 83, 84, 86, 89], and food quality tests [77, 88] (Fig. 1.2a). A disadvantage of colorimetric sensors is the inhomogeneity of the colour distribution and the coffee ring effect in the detection zones, and thus, the judgment of the final colour is challenging by eye [73]. In addition, colorimetric sensors are interfered by the background colour of the paper or the sample. For example, blood assays require

Table 1.2 Rapid diagnostics: sensing mechanisms, their dynamic ranges and sensitivity/ detection limits

Molecular dyes or enzymatic reactions			
Analyte	Dynamic range	Sensitivity/ detection Limit	Reference
Bovine serum albumin	0.38–75 μM	0.75 μM	[70, 71]
Glucose	2.5–500 mM	5 mM	[70, 72–75]
pH	5–9 pH units	0.5 pH units	[73]
Human serum albumin	0.46–46 μM	0.8 μM	[73]
Bendiocarb	<2 μM	~1 nm	[76]
Carbaryl	<2 μM	~10 nm	[76]
Paraoxon	<10 μM	~1 nm	[76]
Malathion	<10 μM	~10 nm	[76]
Aflatoxin B1	<40 μM	~30 nm	[77]
Antibodies to the HIV-1 envelope antigen gp41	1:1–1:100 dilution of HIV in serum	54 fmol/zone (rabbit IgG)	[78]
Lactate	1–25 mM	1 mM	[74]
Uric acid	0.1–7 mM	0.1 mM	[74, 75]
Red cell antigens A, B, and D	Qualitative	N/A	[79, 80]
Urinary acetoacetate	5–16 mM	0.5 mM	[81]
Salivary nitrite	5 μM–2 mM	5 μM	[81]
Total iron	50 μM–1 mM	44 μM	[82]
Hg(II), Ag(I), Cu(II), Cd(II), Pb(II), Cr(VI), Ni(II)	5.4–120.0 ppm	0.001; 0.002; 0.020; 0.020; 0.140; 0.150; 0.230 ppm	[83]
Organic solvents	5–100 % (v/v)	5 % (v/v)	[84]
Hydrogen peroxide	0.1–10 ppm	0.4 ppm	[85]
Volatile organic compounds	Qualitative	Qualitative	[86]
E. coli strain K12 ER2738	$\sim 10^5$– $\sim 10^9$ colony forming unit	$\sim 10^5$ CFU	[87]
E. coli O157:H7, Salmonella Typhimurium, and L. monocytogenes	10–10^3 cfu/cm^2	10 cfu/cm^2	[88]
Particulate metal (Fe)	1.5–15 μg	1.5 μg	[89]
Particulate metal (Ni)	1–15 μg	1 μg	[89]
Particulate metal (Cu)	1–15 μg	1 μg	[89]
Alkaline phosphatase	<500 U/L	~15 U/L	[90]
Aspartate aminotransferase	<300 U/L	~44 U/L	[90]
Aspartate aminotransferase	50–200 U/L	~84 U/L	[91]
Alanine aminotransferase	50–200 U/L	~53 U/L	[91]
Airborne particulate matter	10–100 pmol min^{-1} μg^{-1}	0.32–0.65 ng	[92]
Reactive phosphate	0.2–10 mg L^{-1}	0.05 mg L^{-1}	[93]

(continued)

Table 1.2 (continued)

Molecular dyes or enzymatic reactions

Analyte	Dynamic range	Sensitivity/detection Limit	Reference
Electrochemical sensing			
Glucose	<100 mM	0.21 mM	[94–97]
Lactate	<50 mM	0.36 mM	[94, 95]
Uric acid	<35 mM	1.38 mM	[94, 98]
Ascorbic acid	0.05–0.4 mM	0.02 mmol L^{-1}	[98]
Cholesterol	20–200 mg/dL^{-1}	13 mg/dL^{-1}	[95]
Ethanol	0.1–3 mM	0.2 mM	[95]
Pb(II)	<100 ppb	1.0 ppb	[96, 99, 100]
Cd(II)	10–100 ppb	2.3 ppb	[100]
Cancer and tumour markers (AFP, CA 125-153, CEA)	<100 ng/mL or U/mL	10^{-4} ng mL^{-1}, 3.7×10^{-5} and 2.6×10^{-5} U mL^{-1}, 2.0×10^{-5} ng mL^{-1}	[101–104]
pH	4–10	0.01 pH units	[105]
K$^+$ ions	10^{-5}–10^{-1} M	4.1×10^{-6} M	[105]
NH$_4^+$	10^{-5}–10^{-1} M	7.2×10^{-6} M	[105]
Dopamine	<100 µM	56.6 ± 1.1 mV/unit of pH	[106]
Paracetamol	0.05–2.00 mmol L^{-1}	25 µmol L^{-1}	[107]
4-aminophenol	0.05–2.00 mmol L^{-1}	10 µmol L^{-1}	[107]
1-butanethiol	2–200 µM	0.5 µM	[108]
Nanoparticles			
DNase I	10^{-5}–10^{-1} unit/µL	10^{-5} unit/µL	[109]
Adenosine	<250 µM	11.8 µM	[110]
Human IgG	<5 mg/L	10 µg/L	[111]
Pseudomonas aeruginosa, Staphylococcus aureus	0.5–10×10^3 CFU/mL	500–5,000 CFU/ml	[112]
Glucose	0.5–100 mM	0.5 mM	[113]
Hg(II)	5–75 ppm	0.12 ppm	[114]
*Pf*HRP2	5–40 ng/mL	2.9 ng/mL	[115]
Goat IgG	0.05–1 µg/mL	20 ng mL^{-1}	[116]
Prostate-specific antigen	0.05–100 µg/L	~360.2 ng/L	[117]
Cu^{2+} ions	7.8–62.8 µM	7.8 nM	[118]
HIV DNA	10–10^5 HIV *gag* DNA	10 copies	[119]
Mycobacterium tuberculosis	< 30 µg mL^{-1}	10 µg mL^{-1}	[120]
Immunoglobulin E	0.05–5 pmol	50 fmol	[121]
Cd^{2+} ions	10 ppb	0.1–0.4 ppb	[122]

(continued)

Table 1.2 (continued)

Molecular dyes or enzymatic reactions			
Analyte	Dynamic range	Sensitivity/ detection Limit	Reference
Electrochemiluminescence			
2-(dibutylamino)-ethanol	3 µM–10 mM	0.9 µM	[123]
Nicotinamide adenine dinucleotide	0.2–20.0 mM	72 µM	[123]
Tumour markers (AFP, CA125, CA199, CEA)	5–100 ng or U mL^{-1}	0.15 ng mL^{-1}, 0.6 and 0.17 U mL^{-1}, 0.5 ng mL^{-1}	[103, 124, 125]
Carcinoembryonic antigens	0.01–10 ng mL^{-1}	1 fg mL^{-1}	[126]
Dopamine	1 µM–10 mM	1 µM	[127]
Pb^{2+} ions	<2 µM	10 pM	[128]
Hg^{2+} ions	<2 µM	0.2 nM	[128]
Enzo[a]-pyrene (B[a]P)	0.15–12.5 µM	∼150 nM	[129]
Chemiluminescence			
Glucose	2.5–50 mM L^{-1}	0.14 mmol L^{-1}	[130]
Uric acid	2.5–50 mM L^{-1}	0.52 mmol L^{-1}	[130, 131]
Tumour markers (AFP, CA153, CA199, CEA)	<150 mM ng or U mL^{-1}	1.0 ng mL^{-1}, 0.4, and 0.06 U mL^{-1} and 0.02 ng mL^{-1}	[102, 132]
Fluorescence			
Fluorescein isothiocyanate-labeled bovine serum albumin	0.1–100 pmol	125 fmol	[133]
DNA	10–10,000 pM	100 pM	[134]
Nucleic acids	0.08–5.00 µM	1,200 input templates	[135]
Lung cancer associated miRNA	10 nM–10 µM	∼100 nM	[136]
Nucleic acid hybridisation	1–5 pmol	300 fmol	[137]
Asian soybean rust	0.0032–3.2 mg/mL	2.2 ng/mL	[138]
Dual electrochemical/colorimetric sensing			
Au(III)	1–200 ppm	1 ppm	[139]
Bacterial whole-cell			
Bacterial quorum signalling molecules–N-acylhomoserine lactones	10^{-9}–10^{-4} M	10^8 M	[140]

Table 1.3 The principles of colorimetric reactions in commercial urinalysis tests

Assay	Reaction	Range	Detection limit [144]
Glucose	Glucose + O_2 $\xrightarrow{\text{glucose oxidase}}$ Gluconolactone + H_2O_2 Chromogen + H_2O_2 $\xrightarrow{\text{peroxidase}}$ H_2O + dye	5.5–55	2.2 mmol/L
Protein	Tetrabromophenol blue (pH 3, yellow) $\xrightarrow{\text{protein}}$ tetrabromophenol blue (>pH 4.6, blue)	30–300	6 mg/dL albumin
Blood	haemoglobin H_2O_2 + chromogen $\xrightarrow{\text{peroxide}}$ oxidised chromogen + H_2O_2	10–80	5–10 erythrocytes/L
Leukocytes	Indoxyl or pyrole carbonic acid ester $\xrightarrow[\text{exterase}]{\text{granulocytic}}$ Indoxyl or pyrole Indoxyl or pyrole + diazonium salt → dye (purple)	15–125	10–25 leukocytes/μL
Nitrite	Nitrite + p-arsanilic acid → diazonium compound 3-Hydroxyl-1,2,3,4 tetrahydrobenz-(h)-quinoline + diazonium compound → pink colour	Qualitative	11 μmol/L
pH	HInd (acid form) + H_2O $\rightleftarrows$ H_3O^+ + Ind^- (conjugate base of the indicator)	4–9	0.25 pH units
Ketone	Acetoacetic acid + Na nitroprusside + glycine $\xrightarrow{\text{alkaline}}$ violet to purple colour	0.5–4.0	5 mg/dL
Urobilinogen	p-diethylaminobenzaldehyde + urobilinogen → azo compound (red)	17–200 μmol/L	7 mol/L
Bilirubin	Bilirubin + diazide $\xrightarrow{\text{acid}}$ azobilirubin	+–+++	9 μmol/L

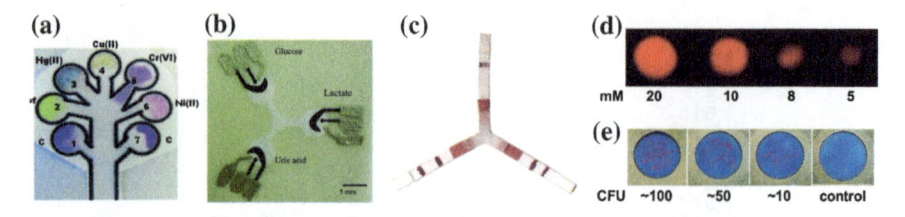

Fig. 1.2 Sensors utilised for point-of-care testing, **a** Colorimetric detection of heavy metals, **b** Electrochemical sensing of glucose, lactate and uric acid, **c** Antibody conjugated gold NP detection of *Pseudomonas aeruginosa* and *Staphylococcus aureus*, **d** ECL emission from a paper-based device at different analyte concentrations, **e** Fluorescent sensing for measuring bacterial growth. Adapted from Ref. [69] with permission from The Royal Society of Chemistry

either usage of serum/plasma samples or a red blood cell filter. Despite its drawbacks, colorimetric sensing is widely used, and it can also be quantified by a handheld reader [141, 142] or a smartphone camera [72, 143].

1.2.2 Electrochemical Sensors

Electrochemistry and ion-selective electrodes have been explored widely due to their well-known principle of operation and maturity. Traditionally, electrochemical sensors have three electrodes: a counter electrode, a working electrode, and a reference electrode. In paper-based assays, carbon ink was used for the counter and the working electrode, whereas silver/silver chloride ink was used for the fabrication of the reference electrode. The reaction zones comprised this multiple-electrode mechanism (Fig. 1.2b). Electrochemical sensors allowed the detection of glucose [94, 95], lactate [94, 95], uric acid [94], cholesterol [95], tumour markers [101], dopamine [106] and drugs [107]. Other studies described heavy metal sensors for environmental monitoring [94, 96, 99, 100]. As compared to colorimetric reagents, the fabrication of electrochemical sensors required an additional deposition step of conductive inks on the paper matrix. All these electrodes and electronic wires were screen printed using graphite and silver inks, respectively. When the sample was introduced to the device, it was wicked up into the sensing zones and the amperometric measurement was initiated via a glucometer. Alternatively, gold can be sputtered on the paper matrix through a shadow mask for the deposition of electrodes [107]. Such electrodes can be characterised by cyclic and square wave voltammetry, and chronoamperometry. In contrast to colorimetric reagents, electrochemical sensors respond within seconds and have sensitivities down to nM [145]. Electrochemical detection is also independent of the ambient light and is less prone to interference from the colour/deteriorations of substrate. Although the attributes of electrochemical detection such as maturity and suitability for miniaturisation are attractive, the requirement for a readout device increases its complexity and the cost per test.

1.2.3 Colloidal Nanoparticles (NPs)

Functionalised colloidal gold and monodisperse latex are the historical sensing reagents for lateral-flow assays, which do not require a readout device for qualitative results [146]. Lateral-flow tests employ antibody-conjugated gold NPs, which have extinction coefficients that are higher than common organic dyes. Inkjet printers have been used to deposit the NPs for multiplex detection on filter paper [111]. Other NP-based assay studies focused on improving the sensitivity of current lateral-flow tests by using paper network platforms. The 2D nitrocellulose-based networks enabled multistep processes to amplify the signal in immunoassays to improve the limit of detection [115, 147]. These cards contained reagents stored dry and the assay was activated by wicking the sample in a single-user step. They were capable of multistep processes such as delivering rinse buffers and signal amplification reagents to the capture zones. These devices were demonstrated by using porous materials such as nitrocellulose and cellulose depending on the sensing application. Other 2D paper networks involved integrating the inlets of a number of lateral flow assays [112] (Fig. 1.2c), adopting folding techniques [119] and microplate paper platforms [120]. NP-based detection has been demonstrated with metabolites [109, 111, 113, 117, 121, 147], bacterial agents [112] in disease diagnosis such as HIV [119], malaria [115], tuberculosis [120] and in environmental monitoring [114, 118]. To multiplex the assay, monodisperse latex can be coupled with fluorescent and coloured dyes, and para/magnetic components. For example, conjugated with dark dyes, monodisperse latex particles exhibit high contrast on nitrocellulose or paper. Additionally, latex particles can be utilised with different colours or fluorescent dyes.

1.2.4 Chemiluminescence (CL)

CL is based on the emission of light generated by a chemical reaction. In the presence of reactants A (luminol) and B (H_2O_2), and a catalyst or excited intermediate (3-aminophthalate), light is produced along with side products. Peroxidase catalyses the oxidation of luminol to 3-aminophthalate, and the decay of the excited state ($\Diamond$) to a lower energy level results in light emission, which can be enhanced by using phenol derivatives such as p-iodophenol. A typical example of CL is the glow stick, which is based on the reaction of peroxide with a phenyl oxalate ester ([A] + [B] $\rightarrow$ [$\Diamond$] $\rightarrow$ [products] + light). In rapid diagnostics, CL has been explored for the detection of glucose and uric acid [130] and tumour markers [132, 148]. Glucose and uric acid assays were based on oxidase reactions coupled with chemiluminescence reactions of a rhodanine derivative with the generated H_2O_2 in an acidic medium [130]. Uric acid was determined through a chemiluminescence reaction between the rhodanine derivative (3-p-nitrylphenyl-5-(40-methyl-20-sulphonophenylazo) rhodanine) and

H_2O_2 [131]. Studies on the detection of tumour markers involved fabricating sandwich-type immunoassays with a typical luminol-H_2O_2 chemiluminescence system catalysed by Ag^0 NPs [132]. Detection of tumour markers was also achieved by sandwich CL-ELISA with antibodies that were covalently immobilised on a chitosan modified paper zone through glutaraldehyde cross-linking [148]. Correlating the concentration of the analyte and the peak intensity of the emitted light allowed quantitative analysis.

1.2.5 Electrochemiluminescence (ECL)

This sensing mechanism is based on luminescence generated by electrochemical reactions. When electrochemically generated intermediates undergo exergonic reactions, they result in an electronically excited state. This state emits light upon relaxation to a lower level state, and therefore it enables readouts without the requirement for a photodetector. ECL has the advantages of both luminescence and electrochemistry. An ECL sensor based on orange luminescence was demonstrated through the detection of 2-(dibutylamino)-ethanol (DBAE) and nicotinamide adenine dinucleotide by readouts of luminescence [123]. The principle of ECL sensor was based on Ru $(bpy)_3^{2+}$ and DBAE. At the electrode, the amine was oxidised and formed a radical cation $[DBAE^\bullet]^+$, followed by deprotonation to create a $DBAE^\bullet$ radical, which reduced Ru $(bpy)_3^{3+}$ to an excited state. Eventually, Ru $(bpy)_3^{2+\bullet}$ emitted light at 620 nm while relaxing to the ground state [149]. This mechanism served as a coreactant that was oxidised solely by the electrode. In ECL, the electrochemical potential initiated and controlled the chemiluminescence reaction. The electrodes were printed using screen printing and adding an ECL active luminophore followed by drying. The substrates were laminated onto a Zensor screen-printed electrode using an office laminator. The cyclic voltammetry of Ru $(bpy)_3^{2+}$ was used to characterise the electrodes. After the lamination step, an incision was made in the laminate layer. After the introduction of the sample to the assay, the potential was stepped from 0 to 1.15 V for a short period to initiate the ECL. The initiation can be achieved by shifting the potential to a level more positive than the oxidation potential of the ruthenium complex. Chronoamperometry was adopted to generate ECL since it provided control over the reaction time (Fig. 1.2d). The ECL readouts were taken with a camera phone housing to block the ambient light. ECL was connected to the mobile phone battery to obtain a short pulse of low voltage. In addition to these drawbacks, ECL required a photomultiplier tube, which was costly in a miniaturised form. The data was analysed based on the red pixel intensity of the ECL emission, which was correlated with a calibration curve, hence the analyte concentration. Other applications to date included sensing tumour markers [103, 124, 126] and ions [128].

1.2.6 Fluorescence

This detection mechanism was first demonstrated on paper microzone plates. Although paper-based plates are known [150–153], they have recently been suggested for quantitative fluorescence measurements. These paper plates require low sample volumes; 12.5 pmol of fluorescein isothiocyanate-labeled bovine serum albumin generated a relative fluorescence of 2,700 ± 850 a.u. [133]. Although concentration sensitivity and mass sensitivity were comparable with plastic plates, the average relative standard deviation for the replicates of all concentrations was higher. This limitation was attributed to light scattering on the cellulose fibres and the influence of the index of refraction between cellulose and air. Although paper-based plates were suggested as alternatives to conventional plastic multiwell plates, their interpretation required a microplate reader. Other notable studies adopting fluorescent sensing included paper strips comprising DNA-conjugated microgels (MG) for DNA detection [134]. Sensing DNA was accomplished by: (i) targeting DNA promoted ligation of a DNA primer to the MG-bound DNA, (ii) rolling circle amplification (RCA) between the primer and a circle DNA, and (iii) hybridisation of the RCA products and a fluorescent DNA probe. Another study reported a portable device for the growth of bacteria or the amplification of bacteriophages. A fluorescent mCherry reporter was used to quantify the growth of bacteria and the concentration of arabinose [87] (Fig. 1.2e). Another paper-based platform involving fluorescent sensing described the use of non-enzymatic nucleic acid circuits based on strand exchange reactions for the detection of target sequences [135]. Overall, although fluorescence sensing brings new capabilities to point-of-care diagnostics, the feasibility requires reduction in cost and miniaturisation of fluorescence readers.

1.2.7 Genetically-Engineered Cells

Bacterial quorum signalling (QS) molecules, N-acyl homoserine lactones (AHLs), have been used as a sensing mechanism [140]. The bacterial cell-based sensing system comprised two main components: (i) AHL-mediated QS regulatory system as recognition elements, and (ii) β-galactosidase as the reporter enzyme. The bacterial cells were inoculated on paper by liquid drying. The paper strip biosensor detected low concentrations (0.1 nM) of AHLs in saliva. The advent of synthetic biology will accelerate the development of whole-cell based biosensors.

1.3 Next Generation Diagnostics

To expand the current capabilities of point-of-care diagnostics, the materials, sensors and readout devices need to evolve. Extending beyond strip tests and lateral-flow design to multiplexed miniaturised assay configurations will put the existing diagnostics in a context that will allow differentiation of conditions with similar symptoms and improve the treatment options. Biofunctional materials and synthesis of reversibly responsive compounds could lead to reusable tests, which might be applicable to conditions where frequent measurements are required. The control over structural design parameters, advances in deposition of materials through printing will play greater roles in the realisation of the new generation diagnostics. Studies on substate-protein/enzyme interactions, surface energy, release characteristics, assay decay will gain momentum in the realisation of diagnostics beyond R&D. Optimisation of capillary flow parameters in lateral-flow devices and advanced microfluidic devices will allow construction of assays with improved control and sensitivity. To date, limited studies in microfluidic platforms have employed unprocessed samples. Incorporation of sample preparation should not be overlooked. Furthermore, the performance of assay after long-term storage, and time-consuming sensor response in colorimetric test require further investigations. Sensing and detection technologies will also evolve. The search for sensing, quantification and readout within a single equipment-free assay will play a greater role in future diagnostics. These attributes may include user-friendly, fool-proof, text/quantity-reporting capabilities and unexplored sensing mechanisms such as paramagnetic particles, quantum dots, coloured latex particles, and genetically engineered whole cells based on synthetic biology, and other novel materials including graphene, plasmonic materials, and printable gratings [154–159]. In improving the sensitivity and providing a user-friendly interface, bioinspired photonic structures, colloidal crystal arrays, diffraction gratings and holographic sensors can offer newer capabilities and readout approaches [160–165]. Some of these sensing and readout mechanisms might not require physically blocking or shaping the substrate for multiplexing. Although significant time has been devoted to quantification with smartphones/handheld readers, equipment-free approaches should not be overlooked. The use of external readers is a barrier for existing assays, yet this requirement will be a greater challenge in adopting newer sensing platforms. The fast-growing mobile phone market in the developing world has made camera phones a potential platform for quantitatively reading diagnostic assays, and this may standardise the readout devices with improved connectivity [166, 167]. Novel approaches towards instrument-free quantification of analytes will be important contributions to the field. Additionally, the trends show that the microfluidic assay formats will be exploited further by the in vitro diagnostics industry [168–170]. All these advances will lead to multiplexed diagnostics that are capable of identifying the specific etiological agent that causes a particular syndrome, which is a goal that has not been achieved yet. Such assays can explore less utilised clinical samples such as tear fluid with contact lens sensors [171].

Existing prototypes need to be transformed into highly reproducible diagnostic devices. Having performance data does not always yield efficacy after deployment. Possible small-scale trials should experiment with the feasibility and cost-effectiveness after scaling up. The ultimate test in the realisation of diagnostics depends on the acceptance from experts in the commercial diagnostics industry. Today, the rapid diagnostics business is based on a standard lateral flow format, involving the use of nitrocellulose as the reaction matrix. The lateral-flow format is the only ubiquitous, universally applicable platform that can be utilised for simple, qualitative, low cost point-of-care applications, while also having enough capability to be functionalised for highly sensitive, fully quantified, multiplexed assays. Hence, the industrial partners are seeking technologies that have improved capabilities, sensitivity, specificity, and compatibility with the existing manufacturing processes. Only low cost is not enough to achieve market penetration in diagnostics. The value of the low-cost and multiplexed diagnostics will be realised upon reaching communities, where they are needed the most.

References

1. Hagist C, Kotlikoff LJ (2009) Who's going broke? Comparing growth in public healthcare expenditure in ten OECD countries. Hacienda Publica Esp 188:55–72
2. Which path will you take? (2007). Pharma 2020: The vision. PricewaterhouseCoopers
3. Medicines and older people implementing medicines-related aspects of the NSF for older people (2001) Department of Health, United Kingdom, 23634
4. Whitesides GM (2009) A lab the size of a postage stamp, TEDX Boston. http://www.ted.com. Accessed 27 Oct 2014
5. Akram MS, Daly R, Vasconcellos FC, Yetisen AK, Hutchings I, Hall EAH (2015) Applications of paper-based diagnostics. In: Castillo-Leon J, Svendsen WE (eds) Lab-on-a-chip devices and micro-total analysis systems. Springer
6. Nkengasong JN, Mesele T, Orloff S, Kebede Y, Fonjungo PN, Timperi R, Birx D (2009) Critical role of developing national strategic plans as a guide to strengthen laboratory health systems in resource-poor settings. Am J Clin Pathol 131(6):852–857. doi:10.1309/AJCPC51 BLOBBPAKC
7. Schull MJ, Banda H, Kathyola D, Fairall L, Martiniuk A, Burciul B, Zwarenstein M, Sodhi S, Thompson S, Joshua M, Mondiwa M, Bateman E (2010) Strengthening health human resources and improving clinical outcomes through an integrated guideline and educational outreach in resource-poor settings: a cluster-randomized trial. Trials 11:118. doi:10.1186/1745-6215-11-118
8. Lee HH, Allain JP (2004) Improving blood safety in resource-poor settings. Vox Sang 87 (Suppl 2):176–179. doi:10.1111/j.1741-6892.2004.00479.x
9. Allegranzi B, Nejad SB, Combescure C, Graafmans W, Attar H, Donaldson L, Pittet D (2011) Burden of endemic health-care-associated infection in developing countries: systematic review and meta-analysis. Lancet 377(9761):228–241. doi:10.1016/S0140-6736 (10)61458-4
10. Urdea M, Penny LA, Olmsted SS, Giovanni MY, Kaspar P, Shepherd A, Wilson P, Dahl CA, Buchsbaum S, Moeller G, Burgess DCH (2006) Requirements for high impact diagnostics in the developing world. Nature 444(Suppl 1):73–79. doi:10.1038/nature05448
11. Newton PN, Green MD, Fernandez FM, Day NPJ, White NJ (2006) Counterfeit anti-infective drugs. Lancet Infect Dis 6(9):602–613. doi:10.1016/S1473-3099(06)70581-3
12. Yager P, Domingo GJ, Gerdes J (2008) Point-of-care diagnostics for global health. Annu Rev Biomed Eng 10:107–144. doi:10.1146/annurev.bioeng.10.061807.160524
13. Mao X, Huang TJ (2012) Microfluidic diagnostics for the developing world. Lab Chip 12 (8):1412–1416. doi:10.1039/c2lc90022j
14. Water quality for ecosystem and human health (2008) 2 edn. GEMS/Water programme office, UN, Burlington, Ontario, Canada

15. Animal health and the millenium development goals (2010) Animal production and health division, Food and Agriculture Organization of the United Nations
16. Aluwong T, Bello M (2010) Emerging diseases and implications for Millennium Development Goals in Africa by 2015-an overview. Vet Ital 46(2):137–145
17. Global strategy for food safety: safer food for better health (2002) (trans: Department FS). Food Safety Department, WHO, Geneva, Switzerland
18. Bhandari N, Mazumder S, Taneja S, Sommerfelt H, Strand TA, Group IES (2012) Effect of implementation of Integrated Management of Neonatal and Childhood Illness (IMNCI) programme on neonatal and infant mortality: cluster randomised controlled trial. BMJ 344: e1634. doi:10.1136/bmj.e1634
19. Mazumder S, Taneja S, Bahl R, Mohan P, Strand TA, Sommerfelt H, Kirkwood BR, Goyal N, Van Den Hombergh H, Martines J, Bhandari N, Integrated Management of N, Childhood Illness Evaluation Study G (2014) Effect of implementation of Integrated Management of Neonatal and Childhood Illness programme on treatment seeking practices for morbidities in infants: cluster randomised trial. BMJ 349:g4988. doi:10.1136/bmj.g4988
20. Butler D (2010) Cholera tightens grip on Haiti. Nature 468(7323):483–484. doi:10.1038/468483a
21. Andrews JR, Basu S (2011) Transmission dynamics and control of cholera in Haiti: an epidemic model. Lancet 377(9773):1248–1255. doi:10.1016/S0140-6736(11)60273-0
22. Sack DA (2011) How many cholera deaths can be averted in Haiti? Lancet 377 (9773):1214–1216. doi:10.1016/S0140-6736(11)60356-5
23. Walton DA, Ivers LC (2011) Responding to cholera in post-earthquake Haiti. N Engl J Med 364(1):3–5. doi:10.1056/NEJMp1012997
24. Zarocostas J (2011) Cholera epidemic in Haiti is blamed on poor sanitation. BMJ 342:d2944. doi:10.1136/bmj.d2944
25. Webster PC (2011) Lack of clean water exacerbates cholera outbreak in Haiti. Can Med Assoc J (journal de l'Association medicale canadienne) 183(2):E83–E84. doi:10.1503/cmaj.109-3764
26. Free AH, Adams EC, Kercher ML, Free HM, Cook MH (1957) Simple specific test for urine glucose. Clin Chem 3(3):163–168
27. Singer JM, Plotz CM (1956) The latex fixation test. I. Application to the serologic diagnosis of rheumatoid arthritis. Am J Med 21(6):888–892. doi:10.1016/0002-9343(56)90103-6
28. Berson SA, Yalow RS (1959) Quantitative aspects of the reaction between insulin and insulin-binding antibody. J Clin Invest 38:1996–2016. doi:10.1172/JCI103979
29. Southern EM (1975) Detection of specific sequences among DNA fragments separated by Gel-Electrophoresis. J Mol Biol 98(3):503. doi:10.1016/S0022-2836(75)80083-0
30. Towbin H, Staehelin T, Gordon J (1979) Electrophoretic transfer of proteins from polyacrylamide gels to nitrocellulose sheets: procedure and some applications. Proc Natl Acad Sci USA 76(9):4350–4354
31. Goldberg DA (1980) Isolation and partial characterization of the drosophila alcohol dehydrogenase gene. Proc Natl Acad Sci USA 77(10):5794–5798
32. Hawkes R, Niday E, Gordon J (1982) A dot-immunobinding assay for monoclonal and other antibodies. Anal Biochem 119(1):142–147
33. Vaitukaitis JL, Braunstein GD, Ross GT (1972) A radioimmunoassay which specifically measures human chorionic gonadotropin in the presence of human luteinizing hormone. Am J Obstet Gynecol 113(6):751–758
34. Swiss Precision Diagnostics (2014) GmbH, Geneva, Switzerland. http://www.swissprecision diagnostics.com. Accessed 27 Oct 2014
35. Chemstrip® Test Strips. (2014) http://www.poc.roche.com. Accessed 27 Oct 2014
36. Multistix® 10 SG Reagent Strips. (2014) http://www.medical.siemens.com. Accessed 27 Oct 2014
37. Medi-test Combi 11 (2014) http://www.mn-net.com. Accessed 27 Oct 2014
38. Arkray AUTION (2014) Sticks. http://www.arkray.co.jp/english. Accessed 27 Oct 2014

39. YD diagnostics URiSCAN Strips (2014) http://www.yd-diagnostics.com. Accessed 27 Oct 2014
40. Chemstrip® Micral Test Strips (2014) http://www.poc.roche.com. Accessed 27 Oct 2014
41. CLINITEK® Microalbumin Reagent Strips (2014) http://www.medical.siemens.com. Accessed 27 Oct 2014
42. American Bio Medica Corporation (2014) (Kinderhook, New York). http://www.abmc.com. Accessed 27 Oct 2014
43. MP Biomedicals (2014) (Santa Ana, CA). http://www.mpbio.com. Accessed 27 Oct 2014
44. Craig Medical Distribution, Inc. (2014) (Vista, CA). http://www.craigmedical.com. Accessed 27 Oct 2014
45. Alfa Scientific Designs, Inc. (2014) (Poway, CA). http://www.alfascientific.com. Accessed 27 Oct 2014
46. Alere Inc. (2014) (Waltham, MA). http://www.alere.com. Accessed 27 Oct 2014
47. Beckton Dickinson (2014) (Franklin Lakes, NJ). http://www.bd.com. Accessed 27 Oct 2014
48. QUIDEL Corporation (2014) (San Diego, CA). http://www.quidel.com. Accessed 27 Oct 2014
49. Beckman Coulter, Inc., Diagnostics Division (2014) (Brea, CA). http://www.beckman coulter.com. Accessed 27 Oct 2014
50. Meridian Bioscience, Inc. (2014) (Cincinnati, OH). http://www.meridianbioscience.com. Accessed 27 Oct 2014
51. Audit Diagnostics (2014) (Carrigtwohill, Ireland). http://www.auditdiagnostics.ie. Accessed 27 Oct 2014
52. Abaxis Inc. (2014) (Union City, CA). http://www.abaxis.com/veterinary/. Accessed 27 Oct 2014
53. Biomedica MedizinProdukte (2014) GmbH & Co KG, Vienna, Austria. http://www.biomedica.co.at. Accessed 27 Oct 2014
54. Henry Schein, Inc. (2014) (Langen, Germany). http://www.henryscheinbrand.com. Accessed 27 Oct 2014
55. Romer Labs, Inc. (2014) (Union, MO). http://www.romerlabs.com. Accessed 27 Oct 2014
56. Neogen Corporation, (2014) (Lansing, MI). http://www.neogen.com. Accessed 27 Oct 2014
57. Charm Sciences, Inc. (2014) (Lawrence, MA). http://www.charm.com. Accessed 27 Oct 2014
58. Kwinbon Biotechnology Co., Ltd. (2014) (Beijing, China). http://www.kwinbon.com. Accessed 27 Oct 2014
59. Wagtech WTD, Palintest Ltd (2014) (Gateshead, U.K.). http://www.wagtech.co.uk. Accessed 27 Oct 2014
60. LaMotte Company (2014) (Chestertown, Maryland). http://www.lamotte.com. Accessed 27 Oct 2014
61. YSI Inc. (2014) (Yellow Springs, Ohio). http://www.ysi.com. Accessed 27 Oct 2014
62. Obreco-Hellige (2014) (Sarasota, FL). http://www.orbeco.com/. Accessed 27 Oct 2014
63. HANNA Instruments US Inc. (2014) (Woonsocket, RI). http://www.hannainst.com. Accessed 27 Oct 2014
64. Industrial Test Systems, Inc. (2014) (Rock Hill, SC). http://www.sensafe.com. Accessed 27 Oct 2014
65. Southern Scientific Ltd (2014) (Henfield, U.K.). http://www.southernscientific.co.uk. Accessed 27 Oct 2014
66. Alexeter Technologies LLC (2014) (Chicago, IL). http://www.alexeter.com. Accessed 27 Oct 2014
67. Miprolab (2014) GmbH (Göttingen, Germany). http://www.miprolab.com. Accessed 27 Oct 2014
68. Wong RC, Tse HY (2009) Lateral flow immunoassay. Humana Press, Springer, New York. doi:10.1007/978-1-59745-240-3
69. Yetisen AK, Akram MS, Lowe CR (2013) Paper-based microfluidic point-of-care diagnostic devices. Lab Chip 13(12):2210–2251. doi:10.1039/c3lc50169h

70. Martinez AW, Phillips ST, Butte MJ, Whitesides GM (2007) Patterned paper as a platform for inexpensive, low-volume, portable bioassays. Angew Chem Int Edit 46(8):1318–1320. doi:10.1002/anie.200603817

71. Zhang AL, Zha Y (2012) Fabrication of paper-based microfluidic device using printed circuit technology. AIP Adv 2:022171. doi:10.1063/1.4733346

72. Martinez AW, Phillips ST, Carrilho E, Thomas SW 3rd, Sindi H, Whitesides GM (2008) Simple telemedicine for developing regions: camera phones and paper-based microfluidic devices for real-time, off-site diagnosis. Anal Chem 80(10):3699–3707. doi:10.1021/ac800112r

73. Abe K, Suzuki K, Citterio D (2008) Inkjet-printed microfluidic multianalyte chemical sensing paper. Anal Chem 80(18):6928–6934. doi:10.1021/ac800604v

74. Dungchai W, Chailapakul O, Henry CS (2010) Use of multiple colorimetric indicators for paper-based microfluidic devices. Anal Chim Acta 674(2):227–233. doi:10.1016/j.aca.2010.06.019

75. Chen X, Chen J, Wang F, Xiang X, Luo M, Ji X, He Z (2012) Determination of glucose and uric acid with bienzyme colorimetry on microfluidic paper-based analysis devices. Biosens Bioelectron 35(1):363–368. doi:10.1016/j.bios.2012.03.018

76. Hossain SM, Luckham RE, McFadden MJ, Brennan JD (2009) Reagentless bidirectional lateral flow bioactive paper sensors for detection of pesticides in beverage and food samples. Anal Chem 81(21):9055–9064. doi:10.1021/ac901714h

77. Hossain SM, Luckham RE, Smith AM, Lebert JM, Davies LM, Pelton RH, Filipe CD, Brennan JD (2009) Development of a bioactive paper sensor for detection of neurotoxins using piezoelectric inkjet printing of sol-gel-derived bioinks. Anal Chem 81(13):5474–5483. doi:10.1021/ac900660p

78. Cheng CM, Martinez AW, Gong J, Mace CR, Phillips ST, Carrilho E, Mirica KA, Whitesides GM (2010) Paper-based ELISA. Angew Chem Int Ed Engl 49(28):4771–4774. doi:10.1002/anie.201001005

79. Khan MS, Thouas G, Shen W, Whyte G, Garnier G (2010) Paper diagnostic for instantaneous blood typing. Anal Chem 82(10):4158–4164. doi:10.1021/ac100341n

80. Al-Tamimi M, Shen W, Zeineddine R, Tran H, Garnier G (2012) Validation of paper-based assay for rapid blood typing. Anal Chem 84(3):1661–1668. doi:10.1021/ac202948t

81. Klasner SA, Price AK, Hoeman KW, Wilson RS, Bell KJ, Culbertson CT (2010) Paper-based microfluidic devices for analysis of clinically relevant analytes present in urine and saliva. Anal Bioanal Chem 397(5):1821–1829. doi:10.1007/s00216-010-3718-4

82. Dungchai W, Chailapakul O, Henry CS (2011) A low-cost, simple, and rapid fabrication method for paper-based microfluidics using wax screen-printing. Analyst 136(1):77–82. doi:10.1039/c0an00406e

83. Hossain SM, Brennan JD (2011) beta-galactosidase-based colorimetric paper sensor for determination of heavy metals. Anal Chem 83(22):8772–8778. doi:10.1021/ac202290d

84. Pumtang S, Siripornnoppakhun W, Sukwattanasinitt M, Ajavakom A (2011) Solvent colorimetric paper-based polydiacetylene sensors from diacetylene lipids. J Colloid Interf Sci 364(2):366–372. doi:10.1016/j.jcis.2011.08.074

85. Xu M, Bunes BR, Zang L (2011) Paper-based vapor detection of hydrogen peroxide: colorimetric sensing with tunable interface. ACS Appl Mater Interfaces 3(3):642–647. doi:10.1021/am1012535

86. Eaidkong T, Mungkarndee R, Phollookin C, Tumcharern G, Sukwattanasinitt M, Wacharasindhu S (2012) Polydiacetylene paper-based colorimetric sensor array for vapor phase detection and identification of volatile organic compounds. J Mater Chem 22 (13):5970–5977. doi:10.1039/C2jm16273c

87. Funes-Huacca M, Wu A, Szepesvari E, Rajendran P, Kwan-Wong N, Razgulin A, Shen Y, Kagira J, Campbell R, Derda R (2012) Portable self-contained cultures for phage and bacteria made of paper and tape. Lab Chip 12(21):4269–4278. doi:10.1039/C2lc40391a

88. Jokerst JC, Adkins JA, Bisha B, Mentele MM, Goodridge LD, Henry CS (2012) Development of a paper-based analytical device for colorimetric detection of select foodborne pathogens. Anal Chem 84(6):2900–2907. doi:10.1021/ac203466y

89. Mentele MM, Cunningham J, Koehler K, Volckens J, Henry CS (2012) Microfluidic paper-based analytical device for particulate metals. Anal Chem 84(10):4474–4480. doi:10.1021/ac300309c

90. Vella SJ, Beattie P, Cademartiri R, Laromaine A, Martinez AW, Phillips ST, Mirica KA, Whitesides GM (2012) Measuring markers of liver function using a micropatterned paper device designed for blood from a fingerstick. Anal Chem 84(6):2883–2891. doi:10.1021/ac203434x

91. Pollock NR, Rolland JP, Kumar S, Beattie PD, Jain S, Noubary F, Wong VL, Pohlmann RA, Ryan US, Whitesides GM (2012) A paper-based multiplexed transaminase test for low-cost, point-of-care liver function testing. Sci Transl Med 4(152):152ra129. doi:10.1126/scitranslmed.3003981

92. Sameenoi Y, Panymeesamer P, Supalakorn N, Koehler K, Chailapakul O, Henry CS, Volckens J (2013) Microfluidic paper-based analytical device for aerosol oxidative activity. Environ Sci Technol 47(2):932–940. doi:10.1021/es304662w

93. Jayawardane BM, McKelvie ID, Kolev SD (2012) A paper-based device for measurement of reactive phosphate in water. Talanta 100:454–460. doi:10.1016/j.talanta.2012.08.021

94. Dungchai W, Chailapakul O, Henry CS (2009) Electrochemical detection for paper-based microfluidics. Anal Chem 81(14):5821–5826. doi:10.1021/ac9007573

95. Nie Z, Deiss F, Liu X, Akbulut O, Whitesides GM (2010) Integration of paper-based microfluidic devices with commercial electrochemical readers. Lab Chip 10(22):3163–3169. doi:10.1039/c0lc00237b

96. Nie Z, Nijhuis CA, Gong J, Chen X, Kumachev A, Martinez AW, Narovlyansky M, Whitesides GM (2010) Electrochemical sensing in paper-based microfluidic devices. Lab Chip 10(4):477–483. doi:10.1039/b917150a

97. Liu H, Crooks RM (2012) Paper-based electrochemical sensing platform with integral battery and electrochromic read-out. Anal Chem 84(5):2528–2532. doi:10.1021/ac203457h

98. Carvalhal RF, Kfouri MS, Piazetta MH, Gobbi AL, Kubota LT (2010) Electrochemical detection in a paper-based separation device. Anal Chem 82(3):1162–1165. doi:10.1021/ac902647r

99. Tan SN, Ge L, Wang W (2010) Paper disk on screen printed electrode for one-step sensing with an internal standard. Anal Chem. doi:10.1021/ac1015062

100. Shi JJ, Tang F, Xing HL, Zheng HX, Bi LH, Wang W (2012) Electrochemical detection of Pb and Cd in paper-based microfluidic devices. J Brazil Chem Soc 23(6):1124–1130

101. Ge S, Ge L, Yan M, Song X, Yu J, Huang J (2012) A disposable paper-based electrochemical sensor with an addressable electrode array for cancer screening. Chem Commun 48(75):9397–9399. doi:10.1039/c2cc34887j

102. Wang P, Ge L, Yan M, Song X, Ge S, Yu J (2012) Paper-based three-dimensional electrochemical immunodevice based on multi-walled carbon nanotubes functionalized paper for sensitive point-of-care testing. Biosens Bioelectron 32(1):238–243. doi:10.1016/j.bios.2011.12.021

103. Wang S, Ge L, Zhang Y, Song X, Li N, Ge S, Yu J (2012) Battery-triggered microfluidic paper-based multiplex electrochemiluminescence immunodevice based on potential-resolution strategy. Lab Chip 12(21):4489–4498. doi:10.1039/c2lc40707h

104. Zang D, Ge L, Yan M, Song X, Yu J (2012) Electrochemical immunoassay on a 3D microfluidic paper-based device. Chem Commun 48(39):4683–4685. doi:10.1039/c2cc16958d

105. Novell M, Parrilla M, Crespo GA, Rius FX, Andrade FJ (2012) Paper-based ion-selective potentiometric sensors. Anal Chem 84(11):4695–4702. doi:10.1021/ac202979j

106. Rattanarat P, Dungchai W, Siangproh W, Chailapakul O, Henry CS (2012) Sodium dodecyl sulfate-modified electrochemical paper-based analytical device for determination of dopamine levels in biological samples. Anal Chim Acta 744:1–7. doi:10.1016/j.aca.2012.07.003

107. Shiroma LY, Santhiago M, Gobbi AL, Kubota LT (2012) Separation and electrochemical detection of paracetamol and 4-aminophenol in a paper-based microfluidic device. Anal Chim Acta 725:44–50. doi:10.1016/j.aca.2012.03.011

108. Dossi N, Toniolo R, Pizzariello A, Carrilho E, Piccin E, Battiston S, Bontempelli G (2012) An electrochemical gas sensor based on paper supported room temperature ionic liquids. Lab Chip 12(1):153–158. doi:10.1039/c1lc20663j

109. Zhao WA, Ali MM, Aguirre SD, Brook MA, Li YF (2008) Paper-based bioassays using gold nanoparticle colorimetric probes. Anal Chem 80(22):8431–8437. doi:10.1021/Ac801008q

110. Liu H, Xiang Y, Lu Y, Crooks RM (2012) Aptamer-based origami paper analytical device for electrochemical detection of adenosine. Angew Chem Int Ed Engl 51(28):6925–6928. doi:10.1002/anie.201202929

111. Abe K, Kotera K, Suzuki K, Citterio D (2010) Inkjet-printed paperfluidic immuno-chemical sensing device. Anal Bioanal Chem 398(2):885–893. doi:10.1007/s00216-010-4011-2

112. Li CZ, Vandenberg K, Prabhulkar S, Zhu X, Schneper L, Methee K, Rosser CJ, Almeide E (2011) Paper based point-of-care testing disc for multiplex whole cell bacteria analysis. Biosens Bioelectron 26(11):4342–4348. doi:10.1016/j.bios.2011.04.035

113. Ornatska M, Sharpe E, Andreescu D, Andreescu S (2011) Paper bioassay based on ceria nanoparticles as colorimetric probes. Anal Chem 83(11):4273–4280. doi:10.1021/ac200697y

114. Apilux A, Siangproh W, Praphairaksit N, Chailapakul O (2012) Simple and rapid colorimetric detection of Hg(II) by a paper-based device using silver nanoplates. Talanta 97:388–394. doi:10.1016/j.talanta.2012.04.050

115. Fu E, Liang T, Spicar-Mihalic P, Houghtaling J, Ramachandran S, Yager P (2012) Two-dimensional paper network format that enables simple multistep assays for use in low-resource settings in the context of malaria antigen detection. Anal Chem 84(10):4574–4579. doi:10.1021/ac300689s

116. Liang J, Wang YY, Liu B (2012) Paper-based fluoroimmunoassay for rapid and sensitive detection of antigen. RSC Adv 2(9):3878–3884. doi:10.1039/C2ra20156a

117. Nie J, Zhang Y, Lin L, Zhou C, Li S, Zhang L, Li J (2012) Low-cost fabrication of paper-based microfluidic devices by one-step plotting. Anal Chem 84(15):6331–6335. doi:10.1021/ac203496c

118. Ratnarathorn N, Chailapakul O, Henry CS, Dungchai W (2012) Simple silver nanoparticle colorimetric sensing for copper by paper-based devices. Talanta 99:552–557. doi:10.1016/j.talanta.2012.06.033

119. Rohrman BA, Richards-Kortum RR (2012) A paper and plastic device for performing recombinase polymerase amplification of HIV DNA. Lab Chip 12(17):3082–3088. doi:10.1039/c2lc40423k

120. Veigas B, Jacob JM, Costa MN, Santos DS, Viveiros M, Inacio J, Martins R, Barquinha P, Fortunato E, Baptista PV (2012) Gold on paper-paper platform for Au-nanoprobe TB detection. Lab Chip 12(22):4802–4808. doi:10.1039/c2lc40739f

121. Szucs J, Gyurcsanyi RE (2012) Towards protein assays on paper platforms with potentiometric detection. Electroanal 24(1):146–152. doi:10.1002/elan.201100522

122. Lopez Marzo AM, Pons J, Blake DA, Merkoci A (2013) All-integrated and highly sensitive paper based device with sample treatment platform for Cd(2+) immunodetection in drinking/tap waters. Anal Chem 85(7):3532–3538. doi:10.1021/ac3034536

123. Delaney JL, Hogan CF, Tian J, Shen W (2011) Electrogenerated chemiluminescence detection in paper-based microfluidic sensors. Anal Chem 83(4):1300–1306. doi:10.1021/ac102392t

124. Ge L, Yan J, Song X, Yan M, Ge S, Yu J (2012) Three-dimensional paper-based electrochemiluminescence immunodevice for multiplexed measurement of biomarkers and point-of-care testing. Biomaterials 33(4):1024–1031. doi:10.1016/j.biomaterials.2011.10.065

125. Li W, Li M, Ge S, Yan M, Huang J, Yu J (2013) Battery-triggered ultrasensitive electrochemiluminescence detection on microfluidic paper-based immunodevice based on dual-signal amplification strategy. Anal Chim Acta 767:66–74. doi:10.1016/j.aca.2012.12.053

126. Yan J, Ge L, Song X, Yan M, Ge S, Yu J (2012) Paper-based electrochemiluminescent 3D immunodevice for lab-on-paper, specific, and sensitive point-of-care testing. Chemistry 18 (16):4938–4945. doi:10.1002/chem.201102855

127. Shi CG, Shan X, Pan ZQ, Xu JJ, Lu C, Bao N, Gu HY (2012) Quantum dot (QD)-modified carbon tape electrodes for reproducible electrochemiluminescence (ECL) emission on a paper-based platform. Anal Chem 84(6):3033–3038. doi:10.1021/ac2033968

128. Zhang M, Ge L, Ge S, Yan M, Yu J, Huang J, Liu S (2013) Three-dimensional paper-based electrochemiluminescence device for simultaneous detection of Pb(2+) and Hg(2+) based on potential-control technique. Biosens Bioelectron 41:544–550. doi:10.1016/j.bios.2012.09.022

129. Mani V, Kadimisetty K, Malla S, Joshi AA, Rusling JF (2013) Paper-based electrochemiluminescent screening for genotoxic activity in the environment. Environ Sci Technol 47(4):1937–1944. doi:10.1021/es304426j

130. Yu J, Ge L, Huang J, Wang S, Ge S (2011) Microfluidic paper-based chemiluminescence biosensor for simultaneous determination of glucose and uric acid. Lab Chip 11 (7):1286–1291. doi:10.1039/c0lc00524j

131. Yu J, Wang S, Ge L, Ge S (2011) A novel chemiluminescence paper microfluidic biosensor based on enzymatic reaction for uric acid determination. Biosens Bioelectron 26 (7):3284–3289. doi:10.1016/j.bios.2010.12.044

132. Ge L, Wang S, Song X, Ge S, Yu J (2012) 3D origami-based multifunction-integrated immunodevice: low-cost and multiplexed sandwich chemiluminescence immunoassay on microfluidic paper-based analytical device. Lab Chip 12(17):3150–3158. doi:10.1039/c2lc40325k

133. Carrilho E, Phillips ST, Vella SJ, Martinez AW, Whitesides GM (2009) Paper microzone plates. Anal Chem 81(15):5990–5998. doi:10.1021/ac900847g

134. Ali MM, Aguirre SD, Xu YQ, Filipe CDM, Pelton R, Li YF (2009) Detection of DNA using bioactive paper strips. Chem Commun 43:6640–6642. doi:10.1039/B911559e

135. Allen PB, Arshad SA, Li B, Chen X, Ellington AD (2012) DNA circuits as amplifiers for the detection of nucleic acids on a paperfluidic platform. Lab Chip 12(16):2951–2958. doi:10.1039/c2lc40373k

136. Yildiz UH, Alagappan P, Liedberg B (2013) Naked eye detection of lung cancer associated miRNA by paper based biosensing platform. Anal Chem 85(2):820–824. doi:10.1021/ac3034008

137. Noor MO, Shahmuradyan A, Krull UJ (2013) Paper-based solid-phase nucleic acid hybridization assay using immobilized quantum dots as donors in fluorescence resonance energy transfer. Anal Chem 85(3):1860–1867. doi:10.1021/ac3032383

138. Miranda BS, Linares EM, Thalhammer S, Kubota LT (2013) Development of a disposable and highly sensitive paper-based immunosensor for early diagnosis of Asian soybean rust. Biosens Bioelectron 45C:123–128. doi:10.1016/j.bios.2013.01.048

139. Apilux A, Dungchai W, Siangproh W, Praphairaksit N, Henry CS, Chailapakul O (2010) Lab-on-paper with dual electrochemical/colorimetric determination of gold and iron. Anal Chem 82(5):1727–1732. doi:10.1021/ac9022555

140. Struss A, Pasini P, Ensor CM, Raut N, Daunert S (2010) Paper strip whole cell biosensors: a portable test for the semiquantitative detection of bacterial quorum signaling molecules. Anal Chem 82(11):4457–4463. doi:10.1021/ac100231a

141. Ellerbee AK, Phillips ST, Siegel AC, Mirica KA, Martinez AW, Striehl P, Jain N, Prentiss M, Whitesides GM (2009) Quantifying colorimetric assays in paper-based microfluidic devices by measuring the transmission of light through paper. Anal Chem 81 (20):8447–8452. doi:10.1021/ac901307q

142. Lee DS, Jeon BG, Ihm C, Park JK, Jung MY (2011) A simple and smart telemedicine device for developing regions: a pocket-sized colorimetric reader. Lab Chip 11(1):120–126. doi:10.1039/c0lc00209g

143. Shen L, Hagen JA, Papautsky I (2012) Point-of-care colorimetric detection with a smartphone. Lab Chip 12(21):4240–4243. doi:10.1039/c2lc40741h

144. Roche (2013) Chemstrip 10 MD package insert, cobas

145. Bakker E, Qin Y (2006) Electrochemical Sensors. Anal Chem 78(12):3965–3984. doi:10. 1021/ac060637m
146. Chandler J, Gurmin T, Robinson N (2000) The place of gold in rapid tests. IVD Technol 6 (2):37–59
147. Fu E, Liang T, Houghtaling J, Ramachandran S, Ramsey SA, Lutz B, Yager P (2011) Enhanced sensitivity of lateral flow tests using a two-dimensional paper network format. Anal Chem 83(20):7941–7946. doi:10.1021/Ac201950g
148. Wang S, Ge L, Song X, Yu J, Ge S, Huang J, Zeng F (2012) Paper-based chemiluminescence ELISA: lab-on-paper based on chitosan modified paper device and wax-screen-printing. Biosens Bioelectron 31(1):212–218. doi:10.1016/j.bios.2011.10.019
149. Xue LL, Guo LH, Qiu B, Lin ZY, Chen GN (2009) Mechanism for inhibition of Ru(bpy)(3) (2 +)/DBAE electrochemiluminescence system by dopamine. Electrochem Commun 11 (8):1579–1582. doi:10.1016/j.elecom.2009.05.059
150. Dieterich K (1902) Testing-paper and method of making same. US Patent 691,249
151. Yagoda H (1937) Applications of confined spot tests in analytical chemistry: preliminary paper. Ind Eng Chem Anal Ed 9(2):79–82. doi:10.1021/ac50106a012
152. Herman Y (1938) Test paper. US Patent 2,129,754 A
153. Johnson JL (1967) Microchemical techniques in solving industrial problems. Microchim Acta 55(4):756–762. doi:10.1007/bf01224400
154. Deng S, Yetisen AK, Jiang K, Butt H (2014) Computational modelling of a graphene Fresnel lens on different substrates. RSC Adv 4(57):30050–30058. doi:10.1039/C4ra03991b
155. Kong X-T, Butt H, Yetisen AK, Kangwanwatana C, Montelongo Y, Deng S, Cruz Vasconcellos FD, Qasim MM, Wilkinson TD, Dai Q (2014) Enhanced reflection from inverse tapered nanocone arrays. Appl Phys Lett 105(5):053108. doi:10.1063/1.4892580
156. Kong XT, Khan AA, Kidambi PR, Deng S, Yetisen AK, Dlubak B, Hiralal P, Montelongo Y, Bowen J, Xavier S, Jiang K, Amaratunga GAJ, Hofmann S, Wilkinson TD, Dai Q, Butt H (2014) Graphene based ultra-thin flat lenses. (under review)
157. Vasconcellos FD, Yetisen AK, Montelongo Y, Butt H, Grigore A, Davidson CAB, Blyth J, Monteiro MJ, Wilkinson TD, Lowe CR (2014) Printable surface holograms via laser ablation. ACS Photonics 1(6):489–495. doi:10.1021/Ph400149m
158. Montelongo Y, Tenorio-Pearl JO, Williams C, Zhang S, Milne WI, Wilkinson TD (2014) Plasmonic nanoparticle scattering for color holograms. Proc Natl Acad Sci U S A 111 (35):12679–12683. doi:10.1073/pnas.1405262111
159. Montelongo Y, Tenorio-Pearl JO, Milne WI, Wilkinson TD (2014) Polarization switchable diffraction based on subwavelength plasmonic nanoantennas. Nano Lett 14(1):294–298. doi:10.1021/nl4039967
160. Yetisen AK, Naydenova I, Vasconcellos FC, Blyth J, Lowe CR (2014) Holographic sensors: three-dimensional analyte-sensitive nanostructures and their applications. Chem Rev 114 (20):10654–10696. doi:10.1021/cr500116a
161. Tsangarides CP, Yetisen AK, da Cruz Vasconcellos F, Montelongo Y, Qasim MM, Wilkinson TD, Lowe CR, Butt H (2014) Computational modelling and characterisation of nanoparticle-based tuneable photonic crystal sensors. RSC Adv 4(21):10454–10461. doi:10. 1039/C3RA47984F
162. Yetisen AK, Butt H, da Cruz Vasconcellos F, Montelongo Y, Davidson CAB, Blyth J, Chan L, Carmody JB, Vignolini S, Steiner U, Baumberg JJ, Wilkinson TD, Lowe CR (2014) Light-directed writing of chemically tunable narrow-band holographic sensors. Adv Opt Mater 2(3):250–254. doi:10.1002/adom.201300375
163. Yetisen AK, Qasim MM, Nosheen S, Wilkinson TD, Lowe CR (2014) Pulsed laser writing of holographic nanosensors. J Mater Chem C 2(18):3569–3576. doi:10.1039/C3tc32507e
164. Yetisen AK, Montelongo Y, da Cruz Vasconcellos F, Martinez-Hurtado JL, Neupane S, Butt H, Qasim MM, Blyth J, Burling K, Carmody JB, Evans M, Wilkinson TD, Kubota LT, Monteiro MJ, Lowe CR (2014) Reusable, robust, and accurate laser-generated photonic nanosensor. Nano Lett 14(6):3587–3593. doi:10.1021/nl5012504

165. Yetisen AK, Montelongo Y, Qasim MM, Butt H, Wilkinson TD, Monteiro MJ, Lowe CR, Yun SH (2014) Nanocrystal bragg grating sensor for colorimetric detection of metal ions. (under review)

166. Yetisen AK, Martinez-Hurtado JL, da Cruz Vasconcellos F, Simsekler MC, Akram, Lowe CR (2014) The regulation of mobile medical applications. Lab Chip 14(5):833–840. doi:10.1039/c3lc51235e

167. Yetisen AK, Martinez-Hurtado JL, Garcia-Melendrez A, Vasconcellos FC, Lowe CR (2014) A smartphone algorithm with inter-phone repeatability for the analysis of colorimetric tests. Sens Actuators, B 196:156–160. doi:10.1016/j.snb.2014.01.077

168. Yetisen AK, Jiang L, Cooper JR, Qin Y, Palanivelu R, Zohar Y (2011) A microsystem-based assay for studying pollen tube guidance in plant reproduction. J Micromech Microeng 21 (5):054018. doi:10.1088/0960-1317/21/5/054018

169. Yetisen AK, Volpatti LR (2014) Patent protection and licensing in microfluidics. Lab Chip 14(13):2217–2225. doi:10.1039/c4lc00399c

170. Volpatti LR, Yetisen AK (2014) Commercialization of microfluidic devices. Trends Biotechnol 32(7):347–350. doi:10.1016/j.tibtech.2014.04.010

171. Farandos NM, Yetisen AK, Monteiro MJ, Lowe CR, Yun SH (2014) Contact lens sensors in ocular diagnostics. Adv Healthc Mater. doi:10.1002/adhm.201400504

Chapter 2
Fundamentals of Holographic Sensing

Optical devices that reversibly respond to external stimuli can provide fast, quantitative, visual colorimetric readouts in real-time. They may consist of bioactive recognition elements that can transmit the signal through a transducer embedded within the system. Responsive photonic structures may have applications in chemical, biological and physical sensors for medical diagnostics, veterinary screening, environmental monitoring, pharmaceutical bioassays, optomechanical sensing and security applications. This chapter provides an overview of the fabrication of optical devices, and highlights holography as a practical approach for the rapid construction of optical sensors that operate in the visible spectrum and near infrared. It begins with describing the fundamentals of holography and origins of holographic sensors. The chapter also explains the principle of operation of these devices and discusses the design parameters that affect the readouts. The principles of laser light interference during sensor fabrication and photochemical patterning are discussed. Furthermore, computational readout simulations of a generic holographic sensor through a finite element method are demonstrated. Studied design parameters include optical effects due to lattice spacing, nanoparticle (NP) size and concentration, number of stacks, their distribution, and lattice deficiencies within the sensor. Computational simulations allow designing holographic sensors with predictive optical characteristics.

2.1 Fabrication of Optical Devices

Optical devices have been fabricated from photonic band-gap materials [1–6]. These materials function through the periodic modulation of the refractive index in a dielectric medium, which allows filtering out and diffracting narrow-band wavelengths. These optical nanostructures control the propagation of light within dielectric media. Their applications include reflective coatings on lenses, pigments in paints and inks, waveguides, reflective mirrors in laser cavities and optical devices [7–10]. Over the last two decades several top-down and bottom-up fabrication techniques have been developed: Layer-by-layer stacking [11, 12], electrochemical etching [13], laser-beam-scanning chemical vapour deposition [14], and holographic lithography [15]. However, rapidly fabricating structures at

© Springer International Publishing Switzerland 2015
A.K. Yetisen, *Holographic Sensors*, Springer Theses,
DOI 10.1007/978-3-319-13584-7_2

approximately half the wavelength of the visible light remained a challenge [16]. Self-assembly approaches have also been demonstrated, including photonic crystalline colloidal arrays [17–21], block copolymers [22], opals and inverse opals [23] and nanocomposites [24]. Bottom-up approaches involved self-assembly of preformed building blocks such as monodisperse colloidal objects into periodic gratings. Such building blocks may be silica (SiO_2), polystyrene microspheres, or block copolymers. The symmetry, lattice constant of the crystal and the index of refraction contrast can be controlled to fabricate ordered photonic structures. For example, block copolymers self-assemble into periodic regions through phase separation of chemically different polymer blocks [25]. In order to achieve visible-light Bragg diffraction, the diameter of the colloids with ranges from 100 to 1 μm may be used to form one, two and three-dimensional photonic structures [26–28]. Self-assembled photonic structures reduce materials of fabrication and lower costs as compared to nano/microfabricated photonic devices.

Stimulus-responsive materials have been incorporated into these photonic devices to induce a change in their lattice constants or spatial symmetry of the crystalline array, and refractive index contrast. For example, refractive-index tuneable oxide materials such as WO_3, VO_2, and $BaTiO_3$ have been incorporated in these matrices to produce photonic structures that are sensitive to electric field or temperature [29]. The crystalline colloidal arrays infiltrated with liquid crystals optically responded to an applied external electric field and an increase in the temperature of the device [30–32]. Numerous fabrication strategies and materials science have been developed to build responsive photonic structures for applications in sensing chemical stimuli, temperature variation, light, electrical and magnetic fields and mechanical forces [33–40]. However, the challenges included limited tuneability, slow turnaround times and hysteresis. Another critical fabrication issue has been the narrow response range due to the limited external stimuli-induced changes in the lattice spacing or the index of refraction. To overcome these challenges, polymer chemistries, new building blocks and tuning mechanisms evolve to create practical approaches. These methods offer potential feasibility for producing diffraction gratings. However, the control over the material selection, patterning ability, angle of diffraction, three-dimensional organisation of diffracting gratings and rapid manufacturing have been limited. To overcome these limitations, generic fabrication approaches have been developed to improve the capabilities of incorporating 3D images with optical tuneability [18, 21]. An emerging platform among these approaches is holography, which allows fabrication of optical sensors with Bragg gratings for applications in the quantification of chemical, biological and physical stimuli [41].

2.2 History of Holography

Holography allows recording three-dimensional images of an object or digital information through the use of a light-sensitive material and laser light, or micro/nanofabrication techniques [42–46]. In 1865, Maxwell had proposed theoretically

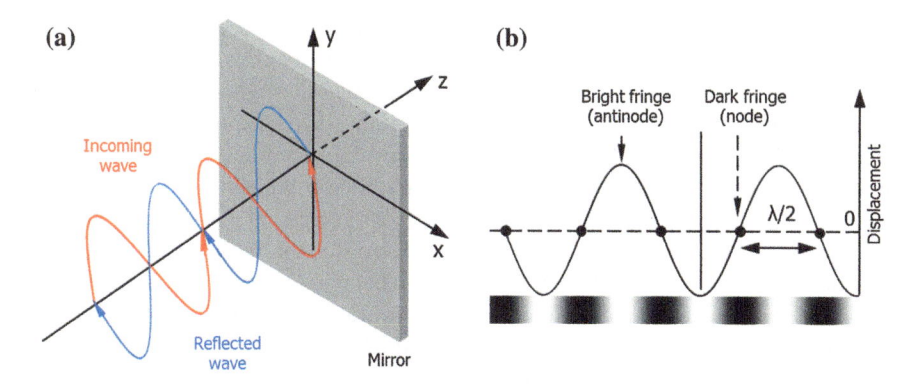

Fig. 2.1 Principle of a standing wave and its corresponding interference pattern. **a** The formation of a standing wave at the plane mirror, **b** Standing wave and interference pattern formed by two coherent beams. Reprinted with permission from [41] Copyright 2014 The American Chemical Society

that light is an electromagnetic disturbance propagating through the field according to electromagnetic laws [47]. In 1869, Zenker theoretically showed that an incident light wave propagating towards a mirror produces a reflected wave, which combines with the incident wave to form an interference pattern with a half-wavelength separation between fringes [48, 49]. In 1887, Hertz experimentally demonstrated the existence of electromagnetic waves by showing that radio waves were consistent with Maxwell's theory [50]. Hertz produced radio standing waves by reflection from a zinc plate. When a monochromatic wave is reflected off a surface, the reflected wave and the incident wave combine to form waves, which oscillate up and down without a direction of motion (Fig. 2.1a). The distance between successive nodes or antinodes is equal to the half of the wavelength of the wave (Fig. 2.1b). Within the standing wave, there is no oscillation at the nodes, while at the antinodes, the oscillations can be maximum.

Zenker's idea to record standing waves of light was experimentally demonstrated by Wiener in the 1890s [51]. He passed carbon arc light, entering a darkroom through a slit, through a prism to filter out the red region of the spectrum. He focused the orthochromatic light using a lens, and directed it perpendicularly to a tilted (2°) 20 nm-thick photographic plate backed by a levelled silver mirror. After he developed and printed the photographic plate, he observed a regular standing wave pattern under magnification. While the antinodes appeared bright, the nodes, containing no light, were dark (Fig. 2.1b). Additionally, the wave might change phase upon reflection, and influence the absolute position of the nodes and antinodes. In 1891, Lippmann developed a method of reproducing colours photographically based on the phenomenon of interference [52]. In his experiment, he projected an image onto a photographic recording medium. The image was produced by shining light through a photographic plate backed by a mirror of liquid mercury, which reflected the light back through the medium to create standing waves. Lippmann was able to create a latent image (an invisible image before development) produced by the standing waves that are characterised by a series of interference maxima and minima. After the recording medium was developed, fixed and dried through the traditional photographic methods, planes of reduced silver particles had reciprocal distances as a function of the wavelength of the light used during recording. Upon illumination with white light, the silver planes diffracted a coloured projection of the recorded image [53]. In the 1910s, X-ray microscopy for recreating the image from the diffraction pattern of a crystal lattice structure were studied by Bragg, Broersch and Zernike [42, 54]. In the 1920s, Wolfke reported that if an X-ray diffraction pattern is illuminated with a monochromatic light, a new diffraction pattern, which is identical with the image of the object is formed [55]. In the late 1940s, Gabor, while trying to improve the resolution of the electron microscope by overcoming the spherical aberration of the lenses, found that adding a coherent background as a phase reference, the original object wave was contained in an interferogram, which he called a hologram [56, 57]. The term hologram was coined from the Greek words *holos*, meaning "whole," and *gramma*, meaning "message". He received the Nobel Prize in physics in 1971 for establishing the principle of holography. However, the stability of the interference required

(i) mechanical and thermal stability of the interferometer used in the holographic recording, and (ii) a coherent light source.

The foundations of the laser (Light Amplification by Stimulated Emission of Radiation) theory were established in the early years of the 20th century by Einstein [58]. In the 1960s, lasers (optical oscillators) were developed to produce monochromatic light [59, 60]. After the development of the laser, Denisyuk of the former Soviet Union, and Leith and Upatnieks in the US recorded independently the first holograms in 1962 [61, 62]. These early holograms were based on silver halide chemistry. Transmission holograms, originally created by Leith and Upatnieks, require monochromatic light (usually a laser) to view the image, otherwise viewing in white light causes severe chromatic aberrations; whereas holograms produced by Denisyuk's method, can be viewed in light of a broad spectral range [43]. Denisyuk was originally inspired by the method of colour photography constructed by Lippmann. In particular, Denisyuk holograms have generated considerable interest in artistic displays, optical devices, data storage and analytical instruments.

Holographic gratings can be generated using various geometries, which involves the use of multiple collimated laser beams. The first step in recording transmission holograms involves passing a single laser beam through a beam splitter, which divides the beam into two beams. The first beam is expanded by a lens, and deviated by mirrors (front surface) onto an object. The light that is scattered back falls onto a recording medium. Meanwhile the second beam, expanded by a lens, travels directly onto the recording medium. The interference of the two mutually coherent beams forms constructive (antinodes) and destructive (nodes) interferences, at regions of the recording medium dictated by the Fourier transform of the object, which implies that all optical information about the object is coded in the

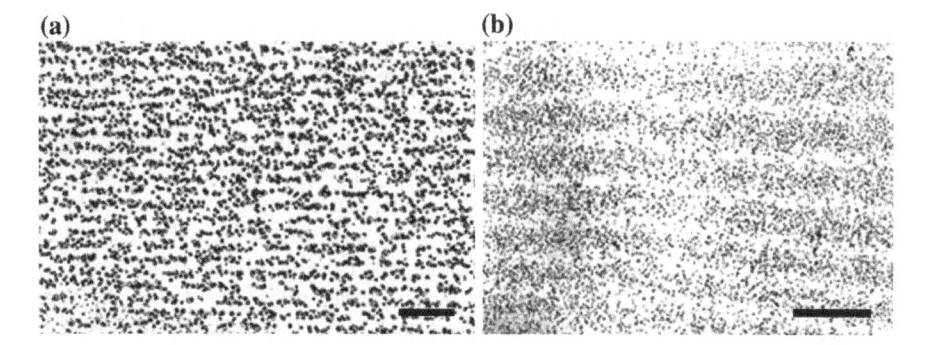

Fig. 2.2 The electron micrographs of hologram cross sections (transversal). **a** A Lippmann phase hologram recorded in a Holotest 8E75HD plate using a HeNe laser operated at 632.8 nm (~50 % diffraction efficiency). Reprinted with permission from [65]. Copyright 1988 The Optical Society of America. **b** A phase hologram recorded in a Slavich PFG-03 M film using a HeNe laser operated at 632.8 nm. *Scale bars* = 1 μm. Reprinted with permission from [41] Copyright 2014 The American Chemical Society

diffraction field produced by the hologram [63]. Holographic recording changes the optical properties of the recording material. An amplitude hologram is recorded when the interference pattern created by the object and the reference beams is copied as variation of the absorption coefficient of the recording material. A phase hologram is created when the holographic recording leads to variation of the refractive index or the thickness of the hologram. Holographic gratings can also be recorded in "Denisyuk" reflection mode. Reflection holograms are typically formed by passing an expanded beam of laser light through the recording plate to illuminate an object on the other side of the plate. Light from the object is then reflected back through the plate and interfered with the light passing through the plate for the first time, thus forming standing waves of light, which are recorded as "holographic fringes" running roughly parallel with the plane of the recording medium [43, 64] (Fig. 2.2). When the hologram is illuminated with a white light source, the fringes in the recording medium act as Bragg mirrors, which diffract light monochromatic (or narrow-band) light and serve as sensitive wavelength filters. The replayed image represents the original object used during the laser exposure. This diffracted light from the periodic gratings results in a narrow-band spectral peak determined by the wavelength of the laser light used and the angle between the two recording beams. The holographic diffraction is governed by Bragg's law:

$$\lambda_{peak} = 2n_0 \Lambda \sin \theta \qquad (2.1)$$

where λ_{peak} is the wavelength of the first order diffracted light at the maximum intensity *in vacuo*, n_0 is the effective index of refraction of the recording medium, Λ is the spacing between the two consecutive recorded NP layers (constant parameter), and θ is the Bragg angle determined by the recording geometry.

2.3 The Origins and Working Principles of Holographic Sensors

In late 1970s, those preparing art holograms realised that so called "pseudo-colour" effects could be obtained by using several exposures of one scene with the 632.8 nm beam from a HeNe laser with each exposure only illuminating different sections of that scene. Before each exposure, the moisture level or pH of the gelatin emulsion was changed, so that each exposure had the emulsion with a different degree of swelling. This resulted in objects having different colours when the finished hologram was replayed under a white light source [66–69]. Specifically, the thickness of the emulsion can be varied through pre-swelling/shrinking the emulsion before laser exposure. Since gelatin film's thickness is greatly affected by its moisture content, moisture control techniques were utilised to create pseudo-colour holograms [68, 70]. In the 1980s, emulsion pre-treatment was optimised to obtain a range of output wavelengths from a fixed exposure wavelength [71, 72]. In the 1990s, the tuning technique of holograms led to the realisation that reflection holograms could be used as sensors to quantify humidity [73] and chemical substances [74–76]. Figure 2.3 shows the timeline in the development of holographic sensors. Any physical or chemical stimulant that changes the lattice spacing (d) or the effective index of refraction (n) of the film cause observable changes in the wavelength (λ_{peak}) or its profile (colour distribution), or the intensity (brightness) of the hologram. The intensity output by the hologram depends on the modulation depth of the refractive index (dark and bright fringes), and the number of planes present in the polymer matrix. Swelling in the polymer matrix increases the distance between NP spacings and produces a red Bragg peak shift, whereas shrinkage in the matrix shifts the peak to shorter wavelengths. The diffraction grating acts as an optical transducer, whose properties are determined by the physical changes in the polymer matrix. For example, when the polymer matrix is functionalised with a receptor comonomer, which has the ability to draw or expel water from the system upon binding, the degree of swelling indirectly represents the concentration of the target analyte. The shift in the Bragg peak can be monitored using a spectrophotometer, and the sensor can be calibrated based on the inputted physical or chemical change. Hence, the same sensor can be optically or visually interpreted to quantify

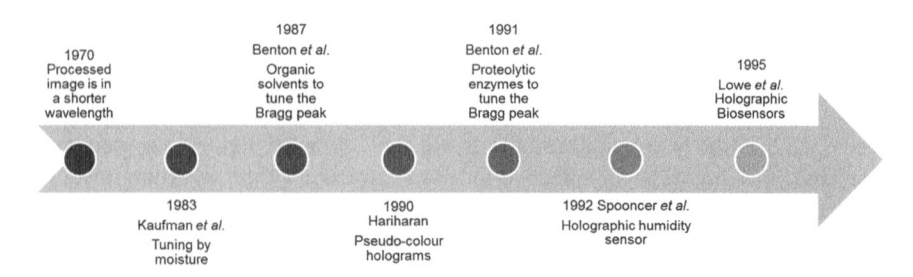

Fig. 2.3 Historical development of holographic sensors

the target analytes in aqueous solutions or physical changes in the environment. This is in contrast to the case of the off-axis transmission hologram, where the holographic fringes run roughly perpendicular to the plane of the plate, and therefore any thickness changes will not greatly shift the Bragg peak. In off-axis transmission gratings, the changes in the brightness or the direction of the diffracted beam can be registered by a photodetector.

Holographic sensors are analytical devices that systematically diffract narrow-band light in the ultraviolet to near-infrared range for application in the detection and quantification of analytes and/or physical parameters [74]. Holographic chemical sensors incorporate gratings within stimuli-responsive polymers, which allow shifting the Bragg peak. Fabrication of the sensors involves laser-directed multi-beam interference and photochemical patterning. The resulting sensors can be interrogated qualitatively by visual inspection or quantitatively by spectrophotom-etry in real-time. The major advantages of holographic sensors over other optical sensors are the ability to produce three-dimensional (3D) images, control over the angle of off-axis diffraction, and amenability to laser manufacturing. They are functionalised with analyte-responsive materials to construct optical sensors for use in testing, where a visual readout and reversibility are required [77]. Holography allows fabrication of disposable sensors that are lightweight for miniaturisation and multiplexing [78]. Holographic sensors offer three capabilities on a single analytical device: (i) Label-free analyte-responsive polymer, (ii) real-time, reversible quanti-fication of the external stimuli, and (iii) three-dimensional image display. Their applications range from in vitro diagnostics to environmental monitoring (Fig. 2.4).

A holographic sensor can change its optical properties when comes into contact with a target analyte. For example, as a result of detection of an analyte, the sensor could change its spectral response and/or its diffraction efficiency, which in visual terms translates into a change of its colour and/or brightness. The diffraction efficiency of a hologram is described as the ratio of the intensities of the diffracted beam divided by the incident beam. This ratio is a quantitative measure of the brightness of the hologram. Depending on the recording mode and optical prop-erties of the recording media, holograms with controllable optical characteristics can be fabricated. Typical holograms include surface holograms, transmission or reflection holograms, phase or amplitude holograms [79]. For example, "Denisyuk" reflection holograms can be used as colorimetric indicators as they can diffract light when they are illuminated with a white light source. The recording and probing of a reflection hologram is shown in Fig. 2.5a, b. In this recording mode, two coherent beams are incident from the opposite sides of the recording medium. For recording of a "Denisyuk" reflection hologram, two plane waves are normally used, and the lattice spacing of the grating can be determined by Bragg's law (Eq. 2.1).

When the beams have the same incident angles, the interference fringes are parallel to the surface of the recording medium, and an unslanted reflection grating is recorded [80]. The incident angles of the two recording beams can also be different, which produces fringes at an angle with respect to the recording medium surface, and the recorded grating will be slanted. In either case, the fringes run along the bisector line of the angle between the two beams. The optical

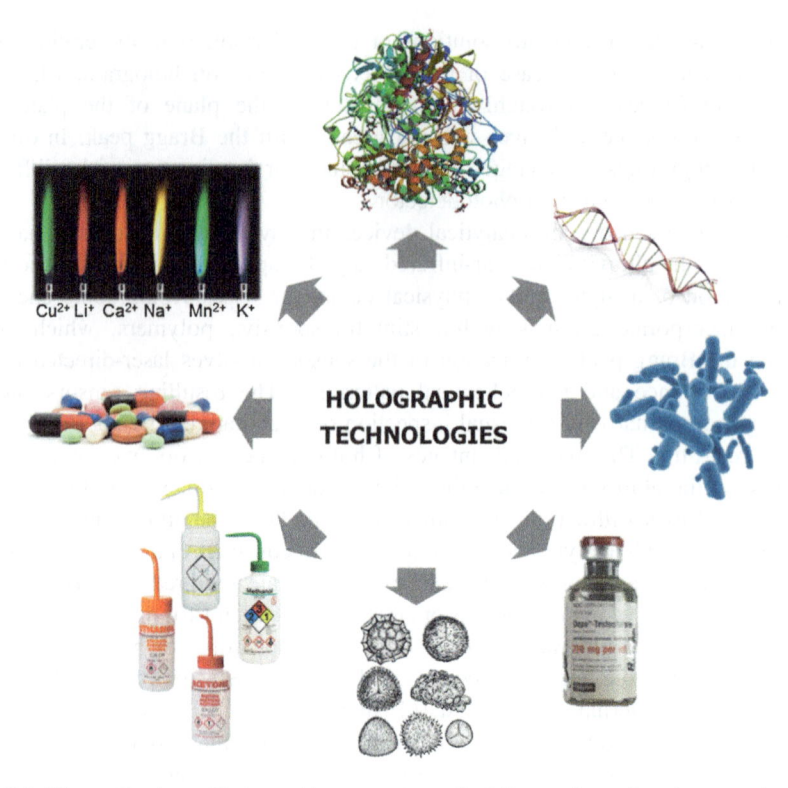

Fig. 2.4 The applications of holographic sensors in medical diagnostics and environmental testing

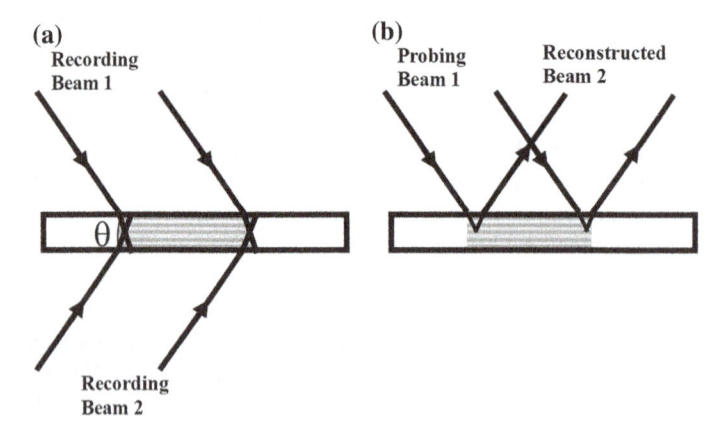

Fig. 2.5 Denisyuk reflection holograms.**a** Recording and **b** probing the hologram. Reprinted with permission from [41] Copyright 2014 The American Chemical Society

characteristic that most conveniently can be observed to change in reflection holograms in the presence of an analyte is the wavelength of the diffracted light. When illuminated with light of a broad spectral range, a reflection hologram diffracts selectively, and operates as a wavelength filter. The maximum diffraction efficiency occurs at a wavelength that satisfies Eq. (2.1). A change in either of the effective refractive index or the lattice spacing causes a change in the wavelength of the diffracted light (Eq. 2.1). It is assumed that the hologram has a thick volume and the angle of observation is constant. In order to quantify how different parameters influence the Bragg peak position, we differentiate Eq. (1.1) using the product rule:

$$\Delta\Lambda = 2\Delta n_0 \Lambda \sin\theta + 2n_0 \Delta\Lambda \sin\theta + 2n_0 \Lambda \cos\theta \Delta\theta \qquad (2.2)$$

where $\Delta\lambda$, Δn_0, $\Delta\Lambda$ and $\Delta\theta$ are the changes in the position of the Bragg peak, effective index of refraction, grating period and the Bragg angle, respectively. We divide both sides of Eq. (2.2) by $2n_0\Lambda \sin\theta$:

$$\frac{\Delta\lambda}{2n_0\Lambda \sin\theta} = \frac{2\Delta n_0 \Lambda \sin\theta}{2n_0\Lambda \sin\theta} + \frac{2n_0 \Delta\Lambda \sin\theta}{2n_0\Lambda \sin\theta} + \frac{2n_0\Lambda \cos\theta \Delta\theta}{2n_0\Lambda \sin\theta} \qquad (2.3)$$

which can be simplified as:

$$\frac{\Delta\lambda}{\lambda} = \frac{\Delta n_0}{n_0} + \frac{\Delta\Lambda}{\Lambda} + \frac{\cos\theta \Delta\theta}{\sin\theta} \qquad (2.4)$$

$$\frac{\Delta\lambda}{\lambda} = \frac{\Delta n_0}{n_0} + \frac{\Delta\Lambda}{\Lambda} + \cot\theta \Delta\theta \qquad (2.5)$$

Using Eq. (2.5), the influence of the changes of optical properties on the Bragg peak can be modelled. Any dimensional change of the polymer matrix in which the hologram is recorded, such as swelling or shrinking produces a change in the lattice spacing, and thus alters the spectral response of the hologram (Fig. 2.6a). A typical Bragg peak shift of a holographic sensor is shown in Fig. 2.6b, and the shift as a function of analyte concentration (Fig. 2.6c). A simulation assuming that the effective refractive index and probe angle remain constant reveals that practically achievable changes in the volume of the polymer matrix could produce large changes in the wavelength of the Bragg peak (Fig. 2.6d). A dimensional change of 30 %, which is normally achieved in an acrylamide-based photopolymer hologram, would produce over a 100 nm shift depending on the initial Bragg peak wavelength [80].

The effective refractive index of the polymer matrix in which the hologram is recorded can change due to the absorption of the target analyte. Assuming that the only property that changes the effective refractive index, the resulting change in the Bragg peak wavelength can be calculated using Eq. (2.5) (Fig. 2.6e). The initial effective refractive index was 1.5. A significant change in the effective refractive index is required in order to obtain a visually observable change in the peak wavelength (Fig. 2.6e). For example, an effective refractive index change of

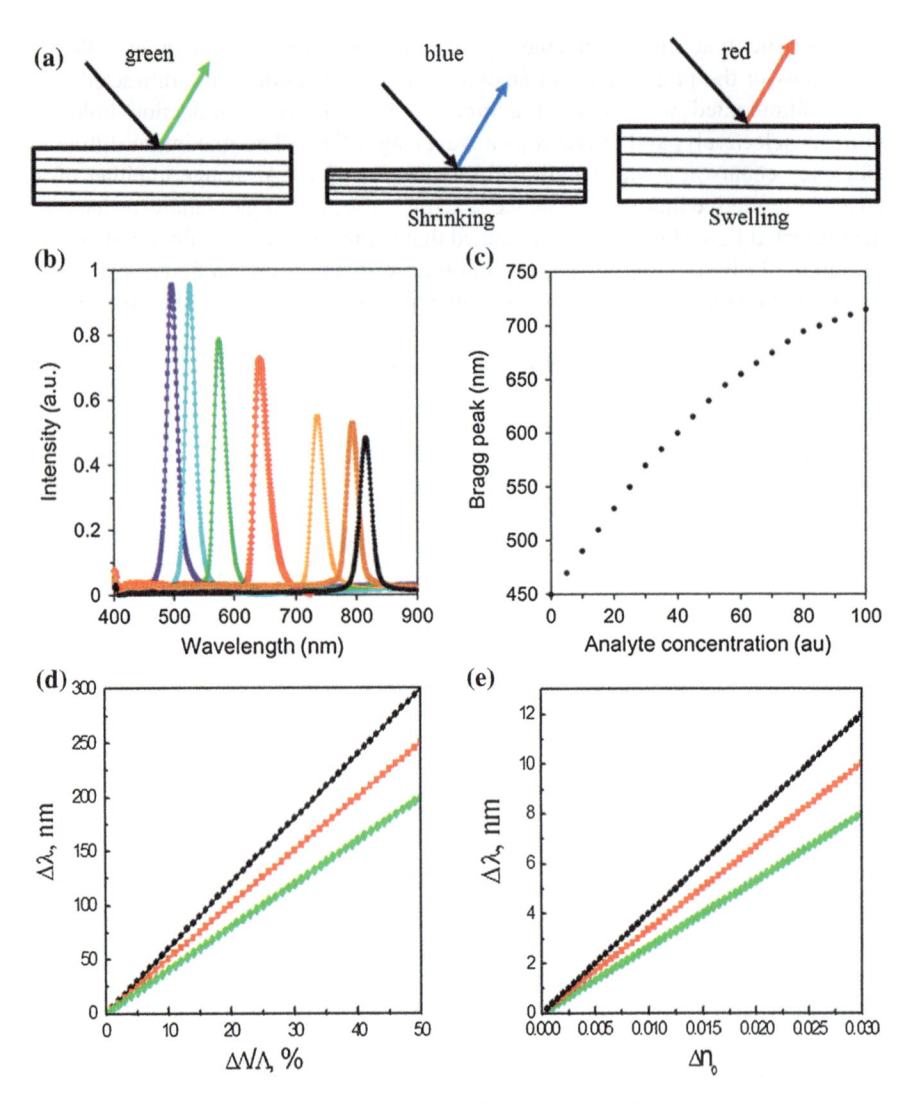

Fig. 2.6 Principle of operation of a Denisyuk reflection holographic sensor. **a** A dimensional change, shrinkage or swelling of the polymer matrix, produces a change in the grating period and a change in the position of the Bragg peak. **b** A typical spectra as the concentration of the analyte changes, **c** The shift in the Bragg peak, **d** Simulated Bragg peak shift as a function of change in lattice spacing. Assuming that the effective refractive index and the Bragg angle (θ) remain constant, as the relative dimensional change ($\Delta\Lambda/\Lambda$) increases, the Bragg peak shifts to longer wavelengths. Practically achievable changes in lattice spacing can produce Bragg shifts larger than 300 nm. The initial Bragg peaks are 600 (●), 500 (■), 400 nm (♦). **e** Bragg peak shift as a function of change in refractive index. Assuming that the initial effective refractive index (n_0) is 1.5 and the Bragg angle (θ) remain constant, as the effective refractive index increases, the Bragg peak shifts to longer wavelengths. Practically achievable changes in refractive index can produce Bragg shifts up to ∼ 12 nm. The initial Bragg peaks are 600 (●), 500 (■), 400 nm (♦). Reprinted with permission from [41] Copyright 2014 The American Chemical Society

15×10^{-3} produces a ~6 nm Bragg peak shift for a sensor originally operating at 600 nm [80]. With the grating period and probe angle remaining constant, it is preferable to record the hologram at a longer wavelength. The absolute change in the peak wavelength ($\Delta\lambda$) can be increased by choosing materials with lower initial effective refractive index (n). Materials with higher porosity have lower effective refractive index. Moreover, for the detection of larger size analytes, it is preferable to use recording media with larger pore size, which allow the diffusion of the analyte into the polymer matrix easily. Both the dimensional and the effective refractive index effects contribute simultaneously to the change in the spectral response of the hologram. For example, in gelatin-based sensors the effective refractive index decreases as the sensor absorbs water and swells; thus, the two factors have opposite contributions to the spectral shift. However, in some materials, one of the factors is the main contributor. For example, in humidity sensors recorded in acrylamide-based photopolymer, the main contributor is the swelling of the polymer matrix due to absorption of moisture studied at relative humidity up to 80 % [81].

2.4 Computational Modelling of Holographic Sensors in Fabrication and Readout

The principles of laser light interference in the fabrication of responsive diffraction gratings are discussed. This chapter is divided into two parts; while the first part explains the photochemical patterning during recording of holographic sensors in Denisyuk reflection mode, the second part describes the operation of the sensors. The first part focuses on the fundamentals of the laser writing in which materials get physically broken, displaced or removed by means of optical forces and thermal energy. In order to understand the different phenomena during photochemical patterning, interference patterns during laser light exposure were simulated. The second part of this chapter demonstrates computational simulations of a generic holographic sensor through a finite element model [82, 83]. To design the sensors with predictive characteristics, its optical properties due to variation in the pattern and the characteristics of the NP arrays were evaluated. Various factors including NP size and distribution within the polymer matrices that directly affect the performance of the sensors were studied computationally.

2.4.1 Photochemical Patterning

In order to predict the interference patterns, which produce the photonic structure in Denisyuk reflection mode, the system was modelled during fabrication as arising from a superposition of different light waves [84–87]. Figure 2.7a shows a schematic of the experimental setup during laser light exposure. Simplifying the simulation of photochemical patterning, the interference pattern created was evaluated from three waves: (1) incident beam (λ_1), (2) beam reflected from the mirror (λ_2), (3) beam reflected internally at the hydrogel-water interface (λ_3). There is a fourth beam that may also be reflected from the mirror, however, this beam was neglected to simplify the simulation. The intensity distribution of the field along the hydrogel was reconstructed by simulating the interference of the three plane waves. Through computing the respective intensities and phases of individual plane waves, the resulting interference pattern was extracted. To visualise the intensity distribution in a 2D cross-section plane, the electromagnetic field in every point over an area of 10×10 μm^2 inside the hydrogel was evaluated. Figure 2.7b shows the interference pattern of the three distinct plane waves, taking into account: the tilt angle, effective index of refraction, laser light wavelength, and exponential decay of laser light intensity while the laser light travels through the hydrogel-Ag^0 NP system and the laser light phase changes upon reflection from the mirror. In order to simulate the laser-induced photochemical patterning, it was assumed that the energy of a single pulse (6 ns, 240 mJ) gets transmitted instantaneously to the particles before heat diffusion is involved. Investigated conditions included materials with considerably thicker than the wavelength of the laser light. Notably, the localisation of heat along the standing wave might be required to produce a well-defined photonic structure. The model was simplified by implying that photochemical patterning occurs, where the energy concentration exceeds a given threshold. Figure 2.7c shows the simulation of the structure after photochemical patterning. In the simulated pattern, black regions correspond to the non-patterned material, while white regions represent photochemically patterned material. Along with the vertical standing wave (~ 193 nm), a larger period wave (~ 3 μm) is in the horizontal direction. Thus, using

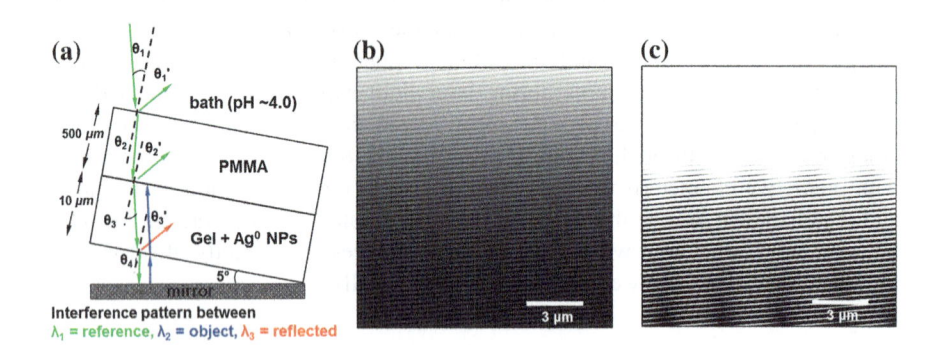

Fig. 2.7 Photochemical patterning based on multi-beam interference in Denisyuk reflection mode. **a** Schematic of the laser light-induced photochemical patterning setup for the preparation of holographic sensors. **b** Intensity field distribution obtained from a holographic sensor with a tilting angle of 5°. The image corresponds to the intensity distribution I = $|E|^2$ produced by laser light interference inside the hydrogel, created by three beams: (1) incident beam, (2) beam reflected from the mirror and (3) beam reflected internally at the pHEMA-water interface. **c** Threshold of intensity of the cross section to represent patterned and nonpatterned regions. The period of the surface grating is 3.01 μm. Reprinted with permission from [84] Copyright 2014 Wiley-VCH Verlag GmbH&Co. KGaA, Weinheim

simulation, we can predict the optical characteristics such as the periodicity of the transmission grating, regions that will be patterned at a given transmission or absorption values and the number of multilayer gratings that will be formed during fabrication.

2.4.2 Simulations of the Optical Readouts

The operation of the holographic sensor is governed by the periodicity of its lattice spacing, which controls the propagation of light through the structure. The lattice periodicity consists of an alternating pattern of mesoscale Ag^0 NP regions organised in a specific direction within a hydrogel [88, 89]. If the absorption of light by the entire structure is minimum and a contrast is present between the periodic Ag^0 NP regions, some frequencies are filtered out as they pass through the photonic structure. The excluded group of frequencies is called the photonic band gap (PBG). The dynamic coloration is generally obtained by altering the periodicity of nano-particle regions either by changing the lattice spacing or the refractive index of the multilayers through chemical reactions. Dynamic coloration in nature include fish (e.g. *Paracheirodon innesi*) [90, 91], cephalopods (e.g. *Euprymna scolopes*) [92] and beetles (e.g. *Tmesisternus isabellae*) [93]. Holographic sensors are analogous to these structures, where the frequency range is designed for a specific PBG. For example, for infrared frequencies, micron dimensions are required for the geometry of the structure [94]. In holographic sensors, the Ag^0 NP-based multilayer structure that was formed within the hydrogel acts as a dynamic 1D photonic crystal, which diffracts the frequencies of electromagnetic radiation that fall within the band gap region. When the band gap region shifts its position to higher or lower frequencies by changing the geometry of the hydrogel, different frequencies are back scattered. To present the principle of operation and provide evidence for subsequent opti-misation of a holographic sensor, a finite element method based on computational software COMSOL Multiphysics®, was utilised [82, 83, 95]. The theoretical diffraction grating consisted of periodic layers of Ag^0 NPs in a hydrogel matrix. The diffraction grating patterns consisting of stacks of randomly-sized Ag^0 NPs were generated using a MATLAB® code. Since the hydrogel matrix has a refractive index close to that of water, and the laser wavelength used for the photochemical patterning was $\lambda = 532$ nm, according to Bragg's law, $\lambda/2n$ results in a lattice constant of $l = 176$ nm. The 1D periodic array of stacks consisted of Ag^0 NPs, which were designed as nanospheres with different radii (Fig. 2.8a).

The simulated geometry consisted of 6 stacks with ~ 60 Ag^0 NPs per stack. Along the vertical axis of each stack, the Ag^0 NPs were uniformly distributed, whilst in the horizontal axis, the Ag^0 NPs were distributed within the layers defined by the laser-induced photochemical patterning. To achieve this, a normal random distribution was performed with the mean positions of the stacks set to a distance equal to the lattice constant. Additionally, to obtain a realistic photonic structure in terms of representing a holographic sensor, a normal random distribution was also

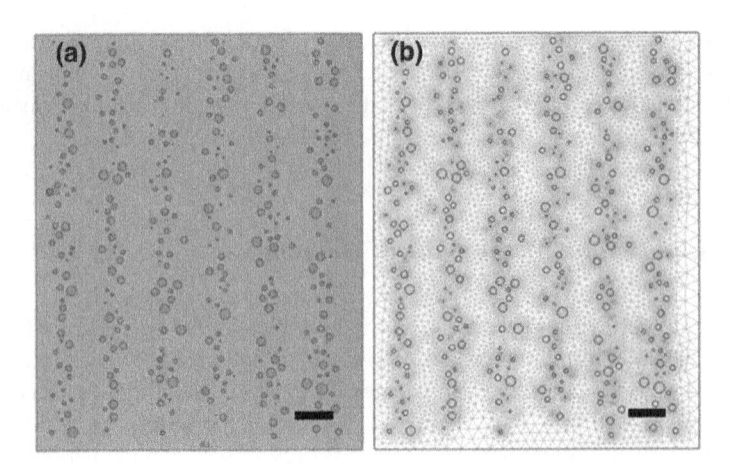

Fig. 2.8 A simulated geometry of a holographic sensor with a multilayer grating. **a** Organisation of Ag^0 NP stacks within a hydrogel matrix, **b** Forming a geometric mesh of the Ag^0 NP pattern. *Scale bars* = 150 nm. Reproduced from [83] with permission from The Royal Society of Chemistry

used to define the radii of the Ag^0 NPs. The mean value of the radii was set to 4–24 nm with $\sigma = 5$ nm. After generating the Ag^0 NP patterns in MATLAB®, they were imported into COMSOL Multiphysics® for modelling. The pattern of Ag^0 NP was surrounded with a square domain of a medium that is analogous to a hydrogel matrix. The remaining Ag^0 NP subdomains were set to have an electrical conductivity of Ag^0 (61.6 mS/m). Since Ag^0 NPs absorbs electromagnetic radiation, a complex refractive index was required. This absorption does not significantly affect the propagation of light when a small number of stacks are simulated. However, the absorption can reduce the efficiency of diffracted light in a holographic sensor that have a high number of Ag^0 NP stacks. Figure 2.8b illustrates the geometric mesh of the holographic sensor in COMSOL Multiphysics®. The incident electromagnetic waves were propagated from left to right along the array of Ag^0 NP stacks. The left boundary of the cell was set to a scattering boundary condition. The light source was defined as a plane wave of varying wavelengths [95]:

$$n \times (\nabla \times H_z) - jkH_z = -jk(1 - k \cdot n)H_{oz}\exp(-jkr) \qquad (2.6)$$

where n is the complex refractive index, H_z is the magnetic field strength at position r, k is the propagation constant, and H_{oz} is the initial magnetic field strength. Meshing was performed with a finite element size of ~ 2 nm to resolve each Ag^0 NP. Once meshing was established, a computation was performed via a parametric sweep, which allowed solving for a range of wavelengths. The wavelength parameters set covered 400–900 nm. Finally, using "power outflow and time average" boundary integration, the transmitted waves were collected at the opposite side of the holographic sensor. Figure 2.9a–c illustrates the simulated geometry that resembles the configuration of a typical holographic sensor, and Fig. 2.9d shows the

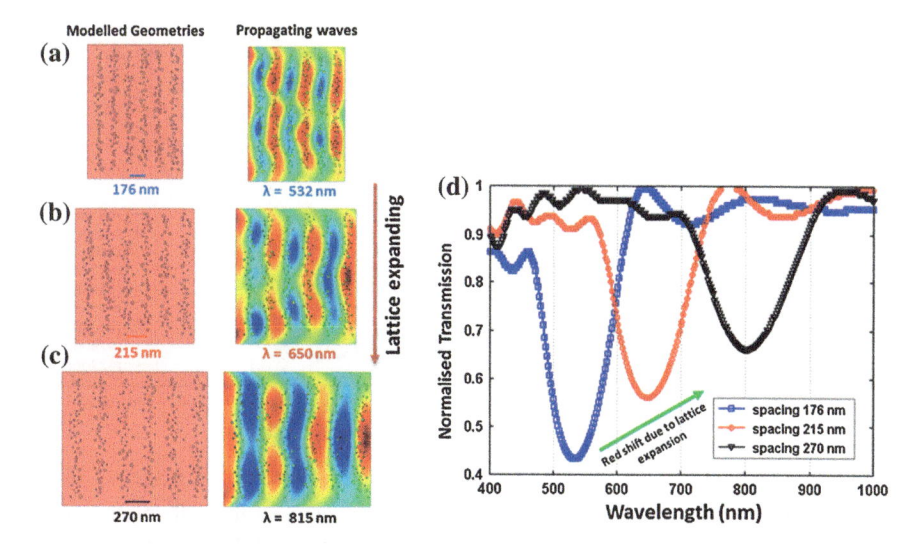

Fig. 2.9 Simulated geometries and transmission spectra for the holographic sensors with lattice constants of **a** 176, **b** 215 and **c** 270 nm. **d** The transmission spectra as a function of lattice spacing. Reproduced from Ref. [83] with permission from The Royal Society of Chemistry

transmission spectra. The spectrum for 176 nm lattice spacing showed peak reflectivity at ~ 532 nm that underwent the lowest transmission due to Bragg diffraction, which defined the diffracted green colour of the holographic sensor (Fig. 2.9a). The colour of the holographic sensor was dictated by the spacing between the Ag^0 NP stacks. The effect of expanding the Ag^0 NP lattice on the reflection band gaps was analysed. The multilayer structure was expanded, while keeping the density and diameter of the Ag^0 NPs constant, simulating the operation of a holographic sensor. The lateral expansion of the polymer matrix increased the effective-stack spacing and stack size, and it reduced the concentration of Ag^0 NPs per stack (Fig. 2.9a–c). The overall effect of these changes on the wave propagation was observed in the simulated transmission spectra, which showed a shift to longer wavelengths in the reflection bands as the Ag^0 NP stack spacing was increased (Fig. 2.9d). The expanding sensor displayed a colour change (reflection band) across the visible spectrum from ~ 532 to 800 nm. As the Ag^0 NP stack spacing increases, the reflection efficiency (intensity) of the sensor decreases. This may be attributed to the decrease in the concentration of Ag^0 NPs present in each stack, which reduces the contrast of the effective refractive index between the Ag^0 NP stacks and the surrounding hydrogel matrix.

The effect of varying Ag^0 NP radii on the efficiency of the holographic sensor was evaluated. Nine different geometries were generated with mean Ag^0 NP radii from 6 to 22 nm, while the number of Ag^0 NPs per stack was kept constant at 60. The transmission spectra showed that as the radii of Ag^0 NPs increased, the intensity of the reflection band also increased, which could be attributed to the area that these respective Ag^0 NPs spread (Fig. 2.10). An increase in the contrast of the

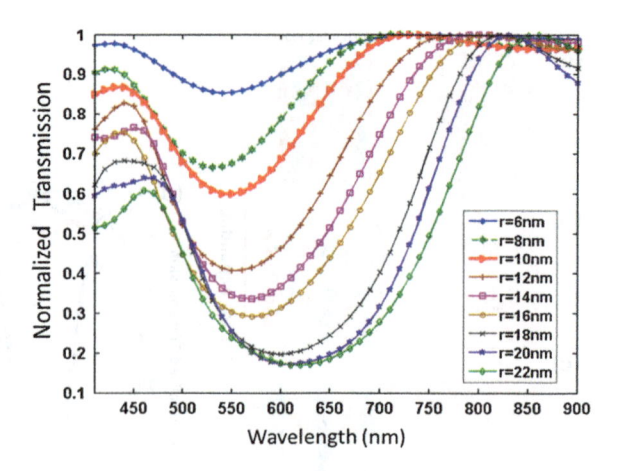

Fig. 2.10 Simulated transmission spectra for the holographic sensors with Ag^0 NPs at different mean radius at a lattice constant of 176 nm with 60 Ag^0 NPs per stack

effective index of refraction of the Ag^0 NP stacks increased the reflection. Therefore, if all these configurations have the same number of Ag^0 NPs, larger Ag^0 NPs would cover more area within the hydrogel matrix, thus resulting in a higher effective index of refraction. However, a photonic structure with Ag^0 NPs larger than ø 100 nm would induce a broad bandwidth and a redshift on the reflection band gap [96]. The broad bandwidth can be explained by the uneven uniformity in the width of the stacks in a holographic sensor. For example, for $r = 22$ nm, not all the five stacks had the same width, because MATLAB® attempted to generate a pattern, where the Ag^0 NPs were evenly spaced inside a stack, and hence it placed them along the horizontal direction. The reflection band shift to longer wavelengths can be explained by the surface plasmonic resonances of the Ag^0 NP [97]. The excitation of surface plasmons arises from a collective electron oscillation within the nanostructure induced by the incident light. This leads to an optical local-field enhancement and a dramatic wavelength-selective photon scattering localised at nanoscale. The plasmonic resonances were affected by the Ag^0 NP radii and geometry, and the optical properties of hydrogel matrix. The reflected light/band gaps displayed by a holographic sensor were influenced by the plasmonic resonances of the Ag^0 NPs. As the radii of the Ag^0 NPs increased, the peak plasmonic resonance shifted to longer wavelengths. Therefore, the band gaps broadened as they represent an effective reflection due to the periodicity of the stacks and the surface plasmon resonances of the larger Ag^0 NPs. The ideal Ag^0 NP radius is between 8 and 10 nm even if they produce weaker reflections than the NP with larger radii. In this radius range, the surface plasmon resonance and the lattice constant dictated band gaps coincide. The reflection efficiency can also be improved by increasing the concentration of the Ag^0 NPs.

The effect of changing the number of Ag^0 NP stacks on the efficiency of the holographic sensor was studied. Three configurations consisting of 60 Ag^0 NPs per stack with a mean radius of 10 nm and a lattice constant of 182 nm were simulated (Fig. 2.11a–c), and their transmission spectra were extracted (Fig. 2.11d).

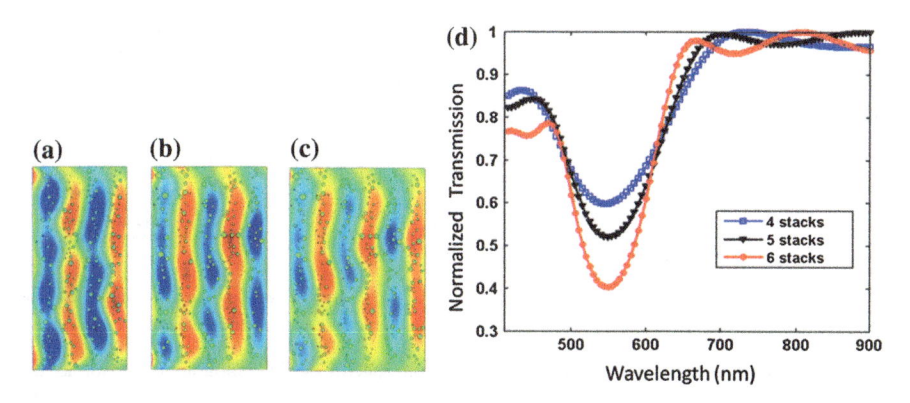

Fig. 2.11 Simulated transmission spectra for the holographic sensors as a function of the number of Ag^0 NP stacks. **a** 4, **b** 5 and **c** 6 stacks of Ag^0 NPs with a lattice constant of 182 nm. **d** The wave propagation spectra for the transmission along the photonic structure with 4, 5 and 5 Ag^0 NP stacks. Stop band is centred at ∼550 nm. Reproduced from [83] with permission from The Royal Society of Chemistry

The position of the reflection band was at 550 nm, and it did not change by adding or removing stacks of Ag^0 NP with the same periodicity. As the number of Ag^0 NP stacks increased, the intensity of the reflected light also increased. For 6, 5, 4 stacks, 60, 48 and 40 % reflection was obtained. At 6 stacks the reflection was stronger than configurations with fewer stacks. Additionally, the lower the reflection, the wider the trough was. The width at half maximum (FWHM) of the 4 stack curve was 160 nm, whereas 6 stacks had 110 nm. Consequently, the greater the number of Ag^0 NP stacks, the deeper the PBG trough and the narrower the bandwidth was.

The concentration of Ag^0 NPs of the holographic sensors was varied from 20 to 80 Ag^0 NPs per stack (Fig. 2.12a–d). Comparing the geometries of the models, as the number of Ag^0 NPs per stack increased, the stacks became more uniform; resembling a continuous medium with fewer voids. Hence, the effective index of refraction of the stacks differed in each case. 20 Ag^0 NPs per stack produced a weak reflection (Fig. 2.12e). With an increasing number of Ag^0 NPs per stack, the reflection band became stronger, with the deepest one reaching 65 % of reflection for 80 Ag^0 NPs per stack. Increasing the number of Ag^0 NP increased the contrast of the index of refraction, thus resulting in higher diffraction efficiencies. However, as the density of Ag^0 NPs per stack increased, the net absorption also increased, leading to effectively lower transmission. The position of the trough shifted to longer wavelengths when the concentration of Ag^0 NPs per stack increased. At 20 Ag^0 NPs per stack, the dip of the curve was located at ∼530 nm, but for 40–60 Ag^0 NPs per stack, the dip was located at ∼555 nm. This shift could be due to a shift in the surface plasmon resonance caused by the close proximity of Ag^0 NPs, and the overall increase in the size of the stacks [96]. Therefore, an increase in the Ag^0 NP concentration per stack resulted in an increase in the refractive index contrast of the holographic sensors.

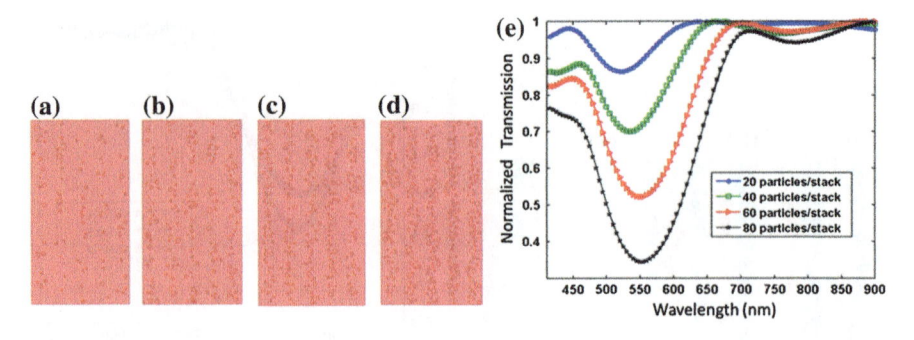

Fig. 2.12 Simulated geometry and the transmission spectra of the holographic sensor as the number of Ag^0 NP stacks was varied. **a** 20, **b** 40, **c** 60 and **d** 80 Ag^0 NPs per stack. **e** The transmission spectra for 20–80 Ag^0 NP per stack. Reproduced from Ref. [83] with permission from The Royal Society of Chemistry

Inhomogeneous Ag^0 NP distribution through the hydrogel matrix may affect the optical properties of the sensor. The effect of anomalies in the holographic sensor was evaluated by simulating four configurations, in which the mean radii of the Ag^0 NPs differed (Fig. 2.13). Distortions are normally present in laser-directed fabrication of holograms, since the Ag^0 NPs are introduced into the polymeric matrices through a diffusion and photographic development, leading to inhomogeneous distribution of NP regions within the hydrogel matrix [98]. The simulated geometries contained six stacks, and they all began with the first stack of Ag^0 NP mean radius of 10 nm. Figure 2.13a–d shows the configurations with Ag^0 NP mean radius that increased by 0.5 nm per stack from 10 to 12 nm. The simulations allowed evaluation of errors due to uncontrolled Ag^0 NP during holographic sensor fabrication. The transmission spectra in Fig. 2.13e show a reference curve for which there is a constant mean radius along all the stacks with the remaining curves representing

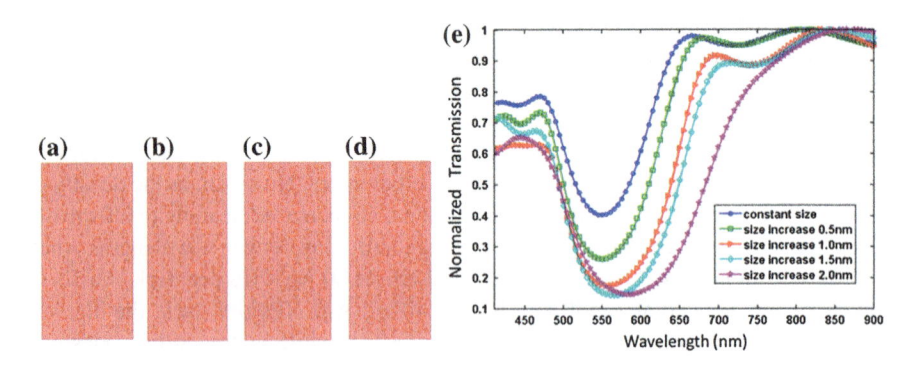

Fig. 2.13 Simulated transmission spectra of the holographic sensor as a function of increasing Ag^0 NP mean radius. Starting from 10 nm, the Ag^0 NP mean radius was increased by **a** 0.5, **b** 1.0, **c** 1.5 and **d** 2.0 nm per stack. **e** Transmission spectra of these configurations as compared to a pattern with constant mean radii. Reproduced from [83] with permission from The Royal Society of Chemistry

a change of mean radius size. In the worst-case scenario of an increase of 2 nm per stack, the curve shows a trough being wider and centred at ~ 585 nm rather than 550 nm. The lattice constant of the reference curve differs from the worst-case scenario, as the latter has a smaller effective lattice constant. In Fig. 2.13d, the distance between the last two stacks is small, hence, the overall effective lattice constant is smaller. Additionally, the spacing between each pair of stacks is non-uniform, which leads to band gap overlaps, which effectively produce a wide bandwidth. This may result in a stronger reflection from the holographic sensor, but poor selectivity (broadband response) for an optical device that needs to exhibit narrow-band peak.

2.5 Conclusions

This chapter described computational modelling of photochemical patterning of recording media in Denisyuk reflection mode, and provided simulations to study the parameters that affect the efficiency of the sensors during readouts. In the first section, grating formation was studied. A threshold of laser light intensity must be passed in order to photochemically pattern the recording media and form a diffraction grating. The produced grating should be produced at the lowest tilt angle to reduce the effect of transmission grating, which broadens the Bragg peak in the readouts. In order to reduce the transmission grating, decane (RI: 1.41) can be used as an index matching fluid during laser exposure, and this inert and odorless compound does not react with hydrogel matrices. In the second part, a photonic multilayer structure based on a stack of Ag^0 NP layers within a hydrogel was computationally studied, and different parameters affecting its performance were evaluated. An optical photonic multilayer structure based on a stack of Ag^0 NP layers within a hydrogel-based system (1D photonic structure) was computationally studied and different parameters affecting its performance were evaluated. The degree of Bragg diffraction and bandwidth of the holographic sensor can be controlled for a desired range by modifying the geometry and distribution of the Ag^0 NP within the hydrogel matrix. The intensity of the reflection band increases as the number of Ag^0 NP stacks increases. The reflection intensity dramatically increases, along with a narrowing of the bandwidth, even by the addition of two extra Ag^0 NP stacks. An increase in the number of Ag^0 NPs is proportional to both the depth and width of the bandwidth. The fabrication of holographic sensors by laser light allows formation of gratings with controlled diffraction angle and pattern [84]. The model demonstrated in this chapter can also be utilised to simulate other nanostructures [99–101]. This theoretical approach allows designing holographic sensors with predictive optical properties, which might reduce the barriers in their integration with point-of-care diagnostic devices [102–105], microfluidic assays [106], contact lens sensors [107] and smartphones [108, 109]. By rationally fabricating holographic sensors with high control over the entire system, including the size and distribution of the Ag^0 NPs within hydrogel matrices, one can avoid undesirable effects such as red-shifted diffraction and wider band gaps.

References

1. Aoki K, Guimard D, Nishioka M, Nomura M, Iwamoto S, Arakawa Y (2008) Coupling of quantum-dot light emission with a three-dimensional photonic-crystal nanocavity. Nat Photonics 2(11):688–692. doi:10.1038/nphoton.2008.202
2. Rinne SA, Garcia-Santamaria F, Braun PV (2008) Embedded cavities and waveguides in three-dimensional silicon photonic crystals. Nat Photonics 2(1):52–56. doi:10.1038/nphoton. 2007.252
3. Takahashi S, Suzuki K, Okano M, Imada M, Nakamori T, Ota Y, Ishizaki K, Noda S (2009) Direct creation of three-dimensional photonic crystals by a top-down approach. Nat Mater 8 (9):721–725. doi:10.1038/nmat2507
4. Ishizaki K, Noda S (2009) Manipulation of photons at the surface of three-dimensional photonic crystals. Nature 460(7253):367–370. doi:10.1038/nature08190
5. Llordes A, Garcia G, Gazquez J, Milliron DJ (2013) Tunable near-infrared and visible-light transmittance in nanocrystal-in-glass composites. Nature 500(7462):323–326. doi:10.1038/ nature12398
6. Kolle M, Lethbridge A, Kreysing M, Baumberg JJ, Aizenberg J, Vukusic P (2013) Bio-inspired band-gap tunable elastic optical multilayer fibers. Adv Mater 25(15):2239–2245. doi:10.1002/adma.201203529
7. Yablonovitch E (2001) Photonic crystals: semiconductors of light. Sci Am 285(6):47–51, 54–45
8. Krauss TF (2003) Photonic crystals—cavities without leaks. Nat Mater 2(12):777–778. doi:10.1038/Nmat1026
9. Akahane Y, Asano T, Song BS, Noda S (2003) High-Q photonic nanocavity in a two-dimensional photonic crystal. Nature 425(6961):944–947. doi:10.1038/nature02063
10. Norris DJ (2007) Photonic crystals. A view of the future. Nat Mater 6(3):177–178. doi:10. 1038/nmat1844
11. Lin SY, Fleming JG, Hetherington DL, Smith BK, Biswas R, Ho KM, Sigalas MM, Zubrzycki W, Kurtz SR, Bur J (1998) A three-dimensional photonic crystal operating at infrared wavelengths. Nature 394(6690):251–253. doi:10.1038/28343
12. Noda S, Tomoda K, Yamamoto N, Chutinan A (2000) Full three-dimensional photonic bandgap crystals at near-infrared wavelengths. Science 289(5479):604–606. doi:10.1126/ science.289.5479.604
13. Birner A, Wehrspohn RB, Gosele UM, Busch K (2001) Silicon-based photonic crystals. Adv Mater 13(6):377–388. doi:10.1002/1521-4095(200103)13:6<377:Aid-Adma377>3.0.Co;2-X
14. Wanke MC, Lehmann O, Muller K, Wen Q, Stuke M (1997) Laser rapid prototyping of photonic band-gap microstructures. Science 275(5304):1284–1286. doi:10.1126/science.275. 5304.1284
15. Campbell M, Sharp DN, Harrison MT, Denning RG, Turberfield AJ (2000) Fabrication of photonic crystals for the visible spectrum by holographic lithography. Nature 404 (6773):53–56. doi:10.1038/35003523
16. Xia YN, Gates B, Yin YD, Lu Y (2000) Monodispersed colloidal spheres: old materials with new applications. Adv Mater 12(10):693–713. doi:10.1002/(Sici)1521-4095(200005)12: 10<693:Aid-Adma693>3.0.Co;2-J
17. Zhao YJ, Zhao XW, Gu ZZ (2010) Photonic crystals in bioassays. Adv Funct Mater 20 (18):2970–2988. doi:10.1002/adfm.201000098

18. Ge J, Yin Y (2011) Responsive photonic crystals. Angew Chem Int Ed 50(7):1492–1522. doi:10.1002/anie.200907091

19. Galisteo-Lopez JF, Ibisate M, Sapienza R, Froufe-Perez LS, Blanco A, Lopez C (2011) Self-assembled photonic structures. Adv Mater 23(1):30–69. doi:10.1002/adma.201000356

20. Buenger D, Topuz F, Groll J (2012) Hydrogels in sensing applications. Prog Polym Sci 37 (12):1678–1719. doi:10.1016/j.progpolymsci.2012.09.001

21. Zhao Y, Xie Z, Gu H, Zhu C, Gu Z (2012) Bio-inspired variable structural color materials. Chem Soc Rev 41(8):3297–3317. doi:10.1039/c2cs15267c

22. Schacher FH, Rupar PA, Manners I (2012) Functional block copolymers: nanostructured materials with emerging applications. Angew Chem Int Ed 51(32):7898–7921. doi:10.1002/anie.201200310

23. Aguirre CI, Reguera E, Stein A (2010) Tunable colors in opals and inverse opal photonic crystals. Adv Funct Mater 20(16):2565–2578. doi:10.1002/adfm.201000143

24. Bonifacio LD, Lotsch BV, Puzzo DP, Scotognella F, Ozin GA (2009) Stacking the nanochemistry deck: structural and compositional diversity in one-dimensional photonic crystals. Adv Mater 21(16):1641–1646. doi:10.1002/adma.200802348

25. Urbas A, Fink Y, Thomas EL (1999) One-dimensionally periodic dielectric reflectors from self-assembled block copolymer-homopolymer blends. Macromolecules 32(14):4748–4750. doi:10.1021/Ma9903207

26. Matsushita SI, Yagi Y, Miwa T, Tryk DA, Koda T, Fujishima A (2000) Light propagation in composite two-dimensional arrays of polystyrene spherical particles. Langmuir 16 (2):636–642. doi:10.1021/La990676b

27. Rogach A, Susha A, Caruso F, Sukhorukov G, Kornowski A, Kershaw S, Mohwald H, Eychmuller A, Weller H (2000) Nano- and microengineering: Three-dimensional colloidal photonic crystals prepared from submicrometer-sized polystyrene latex spheres pre-coated with luminescent polyelectrolyte/nanocrystal shells. Adv Mater 12(5):333–337. doi:10.1002/(Sici)1521-4095(200003)12:5<333:Aid-Adma333>3.0.Co;2-X

28. Yin YD, Lu Y, Xia YN (2001) Assembly of monodispersed spherical colloids into one-dimensional aggregates characterized by well-controlled structures and lengths. J Mater Chem 11(4):987–989. doi:10.1039/B009606g

29. Zhou J, Sun CQ, Pita K, Lam YL, Zhou Y, Ng SL, Kam CH, Li LT, Gui ZL (2001) Thermally tuning of the photonic band gap of SiO_2 colloid-crystal infilled with ferroelectric $BaTiO_3$. Appl Phys Lett 78(5):661–663. doi:10.1063/1.1344574

30. Leonard SW, Mondia JP, van Driel HM, Toader O, John S, Busch K, Birner A, Gosele U, Lehmann V (2000) Tunable two-dimensional photonic crystals using liquid-crystal infiltration. Phys Rev B: Condens Matter Mater Phys 61(4):R2389–R2392. doi:10.1103/PhysRevB.61.R2389

31. Kang D, Maclennan JE, Clark NA, Zakhidov AA, Baughman RH (2001) Electro-optic behavior of liquid-crystal-filled silica opal photonic crystals: effect of liquid-crystal alignment. Phys Rev Lett 86(18):4052–4055. doi:10.1103/PhysRevLett.86.4052

32. Mach P, Wiltzius P, Megens M, Weitz DA, Lin KH, Lubensky TC, Yodh AG (2002) Switchable Bragg diffraction from liquid crystal in colloid-templated structures. Europhys Lett 58(5):679–685. doi:10.1209/epl/i2002-00403-3

33. Gu ZZ, Fujishima A, Sato O (2000) Photochemically tunable colloidal crystals. J Am Chem Soc 122(49):12387–12388. doi:10.1021/Ja005595c

34. Sumioka K, Kayashima H, Tsutsui T (2002) Tuning the optical properties of inverse opal photonic crystals by deformation. Adv Mater 14(18):1284–1286. doi:10.1002/1521-4095 (20020916)14:18<1284:Aid-Adma1284>3.0.Co;2-1

35. Lumsdon SO, Kaler EW, Williams JP, Velev OD (2003) Dielectrophoretic assembly of oriented and switchable two-dimensional photonic crystals. Appl Phys Lett 82(6):949–951. doi:10.1063/1.1541114

36. Fleischhaker F, Arsenault AC, Kitaev V, Peiris FC, von Freymann G, Manners I, Zentel R, Ozin GA (2005) Photochemically and thermally tunable planar defects in colloidal photonic crystals. J Am Chem Soc 127(26):9318–9319. doi:10.1021/ja0521573

37. Xia JQ, Ying YR, Foulger SH (2005) Electric-field-induced rejection wavelength tuning of photonic bandgap composites. Adv Mater 17(20):2463–2467. doi:10.1002/adma.200501166
38. Jeong U, Xia Y (2005) Photonic crystals with thermally switchable stop bands fabricated from Se@Ag2Se spherical colloids. Angew Chem Int Ed 44(20):3099–3103. doi:10.1002/anie.200462906
39. Maurer MK, Lednev IK, Asher SA (2005) Photoswitchable spirobenzopyran-based photochemically controlled photonic crystals. Adv Funct Mater 15(9):1401–1406. doi:10.1002/adfm.200400070
40. Barry RA, Wiltzius P (2006) Humidity-sensing inverse opal hydrogels. Langmuir 22 (3):1369–1374. doi:10.1021/la0519094
41. Yetisen AK, Naydenova I, Vasconcellos FC, Blyth J, Lowe CR (2014) Holographic sensors: three-dimensional analyte-sensitive nanostructures and their applications. Chem Rev 114 (20):10654–10696. doi:10.1021/cr500116a
42. Bjelkhagen HI (1995) Silver-halide recording materials for holography and their processing, 2nd edn. Springer, Heidelberg
43. Saxby G (2004) Practical holography, 3rd edn. Institute of Physics Publishing, London
44. Toal V (2011) Introduction to holography. CRC Press, Boca Raton
45. Bjelkhagen H, Brotherton-Ratcliffe D (2013) Ultra-realistic imaging: advanced techniques in analogue and digital colour holography. Taylor & Francis, Boca Raton
46. Montelongo Y, Tenorio-Pearl JO, Williams C, Zhang S, Milne WI, Wilkinson TD (2014) Plasmonic nanoparticle scattering for color holograms. Proc Natl Acad Sci USA 111 (35):12679–12683. doi:10.1073/pnas.1405262111
47. Maxwell JC (1865) A dynamical theory of the electromagnetic field. Philos Trans R Soc London 155:459–512. doi:10.1098/rstl.1865.0008
48. Zenker W (1868) Lehrbuch der Photochromie (textbook on photochromism). F. Viewag und Suhn, Berlin
49. Guther R (1999) The Berlin scientist and educator Wilhelm Zenker (1829–1899) and the principle of color selection. P Soc Photo-Opt Ins 3738:20–29
50. Hertz H (1893) Electric waves: being researches on the propagation of electric action with finite velocity through space. Macmillan Publishers, London
51. Wiener O (1890) Stehende Lichtwellen und die Schwingungsrichtung polarisirten Lichtes. Ann Phys (Berlin, Ger) 276 (6):203–243. doi:10.1002/andp.18902760603
52. Lippmann G (1894) Sur la Theorie de la Photographie des Couleurs Simples et Composees par la Methode Interferentielle. J Phys 3:97–107
53. Bjelkhagen HI (1997) Lippman photographs recorded in DuPont color photopolymer material, practical holography XI and holographic materials III, vol 3011. SPIE, San Jose
54. Bragg WL (1912) The diffraction of short electromagnetic waves by a crystal. Proc Cambridge Philos Soc 17:43
55. Wolfke M (1920) Über die Möglichkeit der optischen Abbildung von Molekulargittern. Physik Z 21:495–497
56. Gabor D (1948) A new microscopic principle. Nature 161(4098):777–778
57. Gabor D (1949) Microscopy by reconstructed wave-fronts. Proc R Soc A 197(1051):454–487. doi:10.1098/rspa.1949.0075
58. Einstein A (1917) Zur Quantentheorie der Strahlung (On the quantum theory of radiation). Physik Z 18:121–128
59. Gould RG (1959) The LASER, light amplification by stimulated emission of radiation. In: Franken PA, Sands RH (eds) The Ann Arbor conference on optical pumping, Ann Arbor, University of Michigan, p 128
60. Maiman TH (1960) Stimulated optical radiation in ruby. Nature 187(4736):493–494
61. Denisyuk YN (1962) On the reflection of optical properties of an object in a wave field of light scattered by it. Dokl Akad Nauk SSSR 144(6):1275–1278
62. Leith EN, Upatnieks J (1962) Reconstructed wavefronts and communication theory. J Opt Soc Am 52(10):1123–1128. doi:10.1364/JOSA.52.001123
63. Hariharan P (2010) Basics of interferometry. Academic Press, San Diego

64. Benton SA, Bove VM (2007) In-line "Denisyuk" reflection holography. In: Holographic imaging. Wiley, Hoboken, p 173. doi:10.1002/9780470224137.ch16
65. Kubota T (1988) Cross-Sectional view of Lippman hologram gratings. Appl Opt 27 (21):4358–4360. doi:10.1364/AO.27.004358
66. Blyth J (1979) "Pseudoscopic" moldmaking handy trick for Denisyuk holographers. Holosphere 8:5
67. Hariharan P (1980) Pseudocolour images with volume reflection holograms. Opt Commun 35(1):42–44. doi:10.1016/0030-4018(80)90356-9
68. Kaufman JA (1983) Previsualization and pseudo-color image plane reflection holograms. In: Proceeding of international symposium on display holography, Lake Forest College, IL, pp 195–207
69. Moore L (1983) Pseudo-color reflection holography. In: Proceeding of international symposium on display holography, Lake Forest College, IL, p 163
70. Wuest DR, Lakes RS (1991) Color control in reflection holograms by humidity. Appl Opt 30 (17):2363–2367. doi:10.1364/AO.30.002363
71. Walker JL (1987) In situ color control for reflection holography. M.S. Dissertation, MIT, Cambridge, U.S.
72. Walker JL, Benton SA (1989) In-situ swelling for holographic color control. In: Benton SA (ed) Practical holography III, Los Angeles, CA, SPIE, p 192
73. Spooncer RC, Al-Ramadhan FAS, Jones BE (1992) A humidity sensor using a wavelength-dependent holographic filter with fibre optic links. Int J Optoelectron 7(3):449–452
74. Lowe CR, Millington RB, Blyth J, Mayes AG (1995) Hologram used as a sensor. WO Patent Application 1995026499 A1
75. Blyth J, Millington RB, Mayes AG, Frears ER, Lowe CR (1996) Holographic sensor for water in solvents. Anal Chem 68(7):1089–1094. doi:10.1021/ac9509115
76. Millington RB, Mayes AG, Blyth J, Lowe CR (1996) A hologram biosensor for proteases. Sens Actuators B 33(1–3):55–59. doi:10.1016/0925-4005(96)01835-7
77. Postnikov VA, Kraiskii AV, Sergienko VI (2013) Holographic sensors for detection of components in water solutions. In: Mihaylova E (ed) Holography—basic principles and contemporary applications. InTech, Rijeka, p 103. doi:10.5772/53564
78. Mihaylova E, Cody D, Naydenova I, Martin S, Toal V (2013) Research on holographic sensors and novel photopolymers at the centre for industrial and engineering optics. In: Mihaylova E (ed) Holography—basic principles and contemporary applications. InTech, Rijeka, p 89. doi:10.5772/56061
79. Collier RJ, Burckhardt CB, Lin LH (1971) Optical holography. Academic Press, New York
80. Naydenova I, Jallapuram R, Martin S, Toal V (2011) Holographic humidity sensors. In: Okada CT (ed) Humidity sensors: types nanomaterials and environmental monitoring. Nova Science Publishers, Hauppauge, pp 117–142
81. Naydenova I, Jallapuram R, Toal V, Martin S (2009) Characterisation of the humidity and temperature responses of a reflection hologram recorded in acrylamide-based photopolymer. Sens Actuators B 139(1):35–38. doi:10.1016/j.snb.2008.08.020
82. Tsangarides CP (2013) Tuneable photonic crystal-based sensor using silver nanoparticles. MRes in photonic systems development dissertation, University of Cambridge, Cambridge
83. Tsangarides CP, Yetisen AK, da Cruz Vasconcellos F, Montelongo Y, Qasim MM, Wilkinson TD, Lowe CR, Butt H (2014) Computational modelling and characterisation of nanoparticle-based tuneable photonic crystal sensors. RSC Adv 4 (21):10454–10461. doi:10. 1039/C3RA47984F
84. Yetisen AK, Butt H, da Cruz Vasconcellos F, Montelongo Y, Davidson CAB, Blyth J, Chan L, Carmody JB, Vignolini S, Steiner U, Baumberg JJ, Wilkinson TD, Lowe CR (2014) Light-directed writing of chemically tunable narrow-band holographic sensors. Adv Opt Mater 2(3):250–254. doi:10.1002/adom.201300375
85. Yetisen AK, Montelongo Y, da Cruz Vasconcellos F, Martinez-Hurtado JL, Neupane S, Butt H, Qasim MM, Blyth J, Burling K, Carmody JB, Evans M, Wilkinson TD, Kubota LT,

Monteiro MJ, Lowe CR (2014) Reusable, robust, and accurate laser-generated photonic nanosensor. Nano Lett 14(6):3587–3593. doi:10.1021/nl5012504

86. Yetisen AK, Montelongo Y, Qasim MM, Butt H, Wilkinson TD, Monteiro MJ, Lowe CR, Yun SH (2014) Nanocrystal Bragg grating sensor for colorimetric detection of metal ions (under review)

87. Yetisen AK, Qasim MM, Nosheen S, Wilkinson TD, Lowe CR (2014) Pulsed laser writing of holographic nanosensors. J Mater Chem C 2(18):3569–3576. doi:10.1039/C3tc32507e

88. John S (1987) Strong localization of photons in certain disordered dielectric superlattices. Phys Rev Lett 58(23):2486–2489. doi:10.1103/PhysRevLett.58.2486

89. Yablonovitch E (1987) Inhibited spontaneous emission in solid-state physics and electronics. Phys Rev Lett 58(20):2059–2062. doi:10.1103/PhysRevLett.58.2059

90. Mathger LM, Land MF, Siebeck UE, Marshall NJ (2003) Rapid colour changes in multilayer reflecting stripes in the paradise whiptail, Pentapodus paradiseus. J Exp Biol 206 (20):3607–3613. doi:10.1242/Jeb.00599

91. Cong HL, Yu B, Zhao XS (2011) Imitation of variable structural color in Paracheirodon innesi using colloidal crystal films. Opt Express 19(13):12799–12808. doi:10.1364/OE.19.012799

92. Crookes WJ, Ding LL, Huang QL, Kimbell JR, Horwitz J, McFall-Ngai MJ (2004) Reflectins: the unusual proteins of squid reflective tissues. Science 303(5655):235–238. doi:10.1126/science.1091288

93. Liu F, Dong BQ, Liu XH, Zheng YM, Zi J (2009) Structural color change in longhorn beetles Tmesisternus isabellae. Opt Express 17(18):16183–16191. doi:10.1364/OE.17.016183

94. Joannopoulos JD, Johnson SG, Winn JN, Meade RD (2011) Photonic crystals: molding the flow of light, 2nd edn. Princeton University Press, Princeton

95. Zimmerman WBJ (2006) Multiphysics modelling with finite element methods. World Scientific Publishing Company Incorporated, Singapore

96. Tokarev I, Minko S (2012) Tunable plasmonic nanostructures from noble metal nanoparticles and stimuli-responsive polymers. Soft Matter 8(22):5980–5987. doi:10.1039/C2sm25069a

97. Zhang XY, Hu A, Zhang T, Lei W, Xue XJ, Zhou Y, Duley WW (2011) Self-assembly of large-scale and ultrathin silver nanoplate films with tunable plasmon resonance properties. ACS Nano 5(11):9082–9092. doi:10.1021/nn203336m

98. Blyth J, Millington RB, Mayes AG, Lowe CR (1999) A diffusion method for making silver bromide based holographic recording material. Imaging Sci J 47(2):87–91

99. Deng S, Yetisen AK, Jiang K, Butt H (2014) Computational modelling of a graphene Fresnel lens on different substrates. RSC Adv 4(57):30050–30058. doi:10.1039/C4ra03991b

100. Kong X-T, Butt H, Yetisen AK, Kangwanwatana C, Montelongo Y, Deng S, Cruz Vasconcellos FC, Qasim MM, Wilkinson TD, Dai Q (2014) Enhanced reflection from inverse tapered nanocone arrays. Appl Phys Lett 105(5):053108. doi:10.1063/1.4892580

101. Vasconcellos FD, Yetisen AK, Montelongo Y, Butt H, Grigore A, Davidson CAB, Blyth J, Monteiro MJ, Wilkinson TD, Lowe CR (2014) Printable surface holograms via laser ablation. ACS Photonics 1(6):489–495. doi:10.1021/Ph400149m

102. Yetisen AK, Akram MS, Lowe CR (2013) Paper-based microfluidic point-of-care diagnostic devices. Lab Chip 13(12):2210–2251. doi:10.1039/c3lc50169h

103. Volpatti LR, Yetisen AK (2014) Commercialization of microfluidic devices. Trends Biotechnol 32(7):347–350. doi:10.1016/j.tibtech.2014.04.010

104. Yetisen AK, Volpatti LR (2014) Patent protection and licensing in microfluidics. Lab Chip 14(13):2217–2225. doi:10.1039/c4lc00399c

105. Akram MS, Daly R, Vasconcellos FC, Yetisen AK, Hutchings I, Hall EAH (2015) Applications of paper-based diagnostics. In: Castillo-Leon J, Svendsen WE (eds) Lab-on-a-chip devices and micro-total analysis systems. Springer, Berlin. doi:10.1007/978-3-319-08687-3_7

106. Yetisen AK, Jiang L, Cooper JR, Qin Y, Palanivelu R, Zohar Y (2011) A microsystem-based assay for studying pollen tube guidance in plant reproduction. J Micromech Microeng 21(5):054018. doi:10.1088/0960-1317/21/5/054018

107. Farandos NM, Yetisen AK, Monteiro MJ, Lowe CR, Yun SH (2014) Contact lens sensors in ocular diagnostics. Adv Health Mater. doi:10.1002/adhm.201400504

108. Yetisen AK, Martinez-Hurtado JL, da Cruz Vasconcellos F, Simsekler MC, Akram MS, Lowe CR (2014) The regulation of mobile medical applications. Lab Chip 14(5):833–840. doi:10.1039/c3lc51235e

109. Yetisen AK, Martinez-Hurtado JL, Garcia-Melendrez A, Vasconcellos FC, Lowe CR (2014) A smartphone algorithm with inter-phone repeatability for the analysis of colorimetric tests. Sens Actuators B 196:156–160. doi:10.1016/j.snb.2014.01.077

Chapter 3
Holographic pH Sensors

Dynamic photonic structures can be modulated by changing the periodic structure and/or the index of refraction [1–3]. These dynamic photonic structures allow responsive capability for sensing external stimuli to control the properties of light and act as optical transducers [4, 5]. Dynamic optical systems operating in the visible and near-infrared region offer promise for designing adaptive materials and sensors. Such devices have been prepared by microfabrication, self-assembly or a combination of both [6–16]. However, achieving the attributes of a narrow-band response with a wide operating range to construct optical sensors in a few steps in hydrophilic polymers still remains a challenge. This chapter describes the construction of holographic pH sensors by silver chemistry and laser ablation induced in situ size reduction of Ag^0 NPs in hydrogel matrices using Denisyuk reflection holography [17–19]. The holographic sensor consists of chemical-stimuli responsive hydrogels with reversible narrow-band tuneability using Ag^0 NPs that are organised in density-concentrated 3D regions. The optical characteristics of these sensors were investigated by analysing the distribution of the mean diameter of Ag^0 NPs, effective refractive indices of patterned polymer-NP regions, and angular-resolved measurements. The clinical utility of the sensor for the quantification of pH in artificial urine was demonstrated. The chapter also shows strategies for fabricating holographic flakes and paper-based holographic sensors.

3.1 Holographic pH Sensors via Silver-Halide Chemistry

The first step in the fabrication of holographic sensors is the preparation of a substrate, where the hydrogel matrix is deposited. 3-(Trimethoxysilyl)propyl methacrylate in acetone (1:50, v/v) was poured onto glass microscope slides in an aluminium tray [17, 20]. After thorough coating, the excessive acetone/silane mix was poured off whilst slides remained in situ due to surface tension. The slides stored in the tray overnight in the dark, before removal and dark storage at room temperature (Fig. 3.1a). A monomer solution consisting of HEMA (91.5 mol%), EDMA (2.5 mol%) and MAA (6 mol%) was prepared [17, 18, 20]. The solution

© Springer International Publishing Switzerland 2015
A.K. Yetisen, *Holographic Sensors*, Springer Theses,
DOI 10.1007/978-3-319-13584-7_3

was mixed 1:1 (v/v) with DMPA in propan-2-ol (2 %, w/v). An Al coated polyester sheet, Al side facing up, was placed on a levelled surface, and the solution (200 µl) was pipetted as a lateral blob on the surface. Silanised side facing down, the slide was gradually lowered on the lateral blob. Bubbles were prevented by lowering one side of the slide over the blob, then laying the rest of the slide on the lateral blob. The system was exposed to UV light (~ 350 nm) for 1 h (Fig. 3.1b). The slide was peeled off from the aluminium coated sheet, and finally rinsed with ethanol and DI water, respectively (Fig. 3.1c). Silver perchlorate ($AgClO_4$) (0.3 M, 200 µl) dissolved in propan-2-ol and DI water (1:1, v/v) was pipetted as an elongated blob onto a glass sheet. Poly(2-hydroxyethyl methacrylate) (pHEMA) matrix side down, the slide was placed on the elongated blob: The slide remained in situ for 3 min to allow diffusion of $AgClO_4$ into the pHEMA matrix (Fig. 3.1d). The slide was lifted off and excess $AgClO_4$ solution was wiped with a squeegee. The pHEMA matrix was dried under a tepid air current for 5 s (Fig. 3.1e). A solution consisting of 1,1'-diethyl-2,2'-cyanine iodide (1:500, w/v, 1 ml) in methanol and lithium bromide (0.3 M, 40 ml) in methanol:H_2O (3:2, v/v) was dispensed into a Petri dish. pHEMA matrix side facing up, the slide was immersed in the Petri dish for 30 s (Fig. 3.1f). The slide was washed thoroughly with DI water for 30 s (Fig. 3.1g).

After the photosensitisation of the pHEMA matrix, a latent image can be recorded. A levelled Petri dish, mirror in the bottom surface, was filled with unbuffered ascorbic acid (2 % w/v, pH 2.66). pHEMA side facing down, the slide was immersed with an inclination of 5° in the bath. The pHEMA matrix was equilibrated in the bath for 15 min. The matrix was exposed to a single 6 ns pulse, Q-switch delay set to 275 µs by Nd-Yttrium-Aluminum-Garnet (Nd:YAG) pulsed laser ($\lambda = 532$ nm) with a spot size larger than a microscope slide (Fig. 3.1h). A photographic developer (pH ~ 13.0) comprising of 4-methylaminophenol sulphate (0.3 %, w/v), ascorbic acid (2 %, w/v), sodium carbonate (Na_2CO_3) (5 %, w/v) and NaOH (1.5 %, w/v) was dissolved in DI water [21]. The pHEMA matrix was submerged in the developer until no more darkening was seen (Fig. 3.1i). The pHEMA matrix was washed thoroughly with DI water and immersed in acetic acid solution (5 %, v/v) to neutralise the developer (Fig. 3.1j). The pHEMA matrix was rinsed with DI water and immersed in a sodium thiosulphate ($Na_2S_2O_3$) solution (10 %, w/v) mixed with ethanol 1:1 (v/v) for 15 min to remove the undeveloped LiBr crystals (Fig. 3.1k). Finally, the pHEMA matrix was submerged in an ethanol-water (50 % v/v) solution for 15 min to remove the cyanine dye from the polymer matrix and this process was repeated three times (Fig. 3.1l). The holographic sensor can be bleached, which increases the diffraction efficiency. The pHEMA matrix was rinsed with DI water and was immersed in a rehalogenating bleach solution consisting of ethylenediaminetetraacetic acid iron(III) sodium salt hydrate (40 g), potassium bromide, acetic acid (4:6:7, w/w/v) in DI water. The slide was removed after 30 s of bleaching and rinsed with DI water. The bleach converted Ag° NPs to AgBr nanocrystals, yielding an aggregation of AgBr salts in the adjacent dark fringes (constructive interference sites).

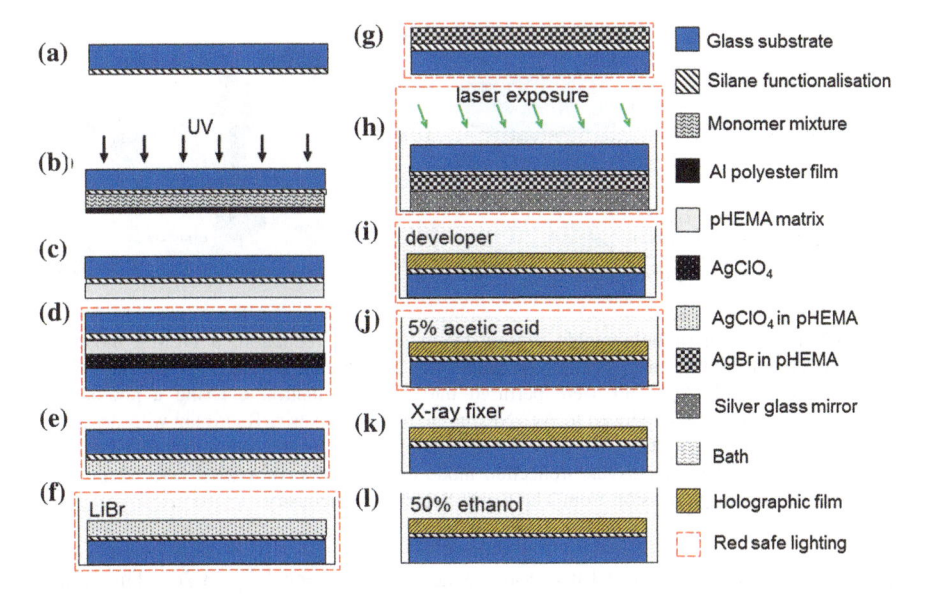

Fig. 3.1 Fabrication of holographic sensors through silver halide chemistry. **a** A glass slide was functionalised with silane chemistry. **b** The monomer mixture was polymerised on the glass slide. **c** The resulting system was rinsed with ethanol. **d** $AgClO_4$ solution was allowed to diffuse into the pHEMA matrix. **e** The excess $AgClO_4$ solution was removed and the pHEMA matrix was dried. **f** AgBr nanocrystals were formed in the pHEMA matrix. **g** The matrix was rinsed with DI water. **h** The pHEMA matrix was exposed to a single pulse of a laser light at 5°. **i** The latent image sites were developed to silver metal (Ag^0). **j** The pHEMA matrix was neutralised. **k** Undeveloped AgBr grains were removed from the pHEMA matrix using hypo. **l** The pHEMA matrix was rinsed with ethanol solution in order to remove the cyanine dye from the matrix. Adapted from Ref. [17] with permission from The Royal Society of Chemistry

3.2 Fabrication of Holographic pH Sensors Through in Situ Size Reduction of Ag^0 NPs

Depending on its manufacturing technique, polymethyl (methacrylate) (PMMA) may have comparable optical properties to glass [22, 23]. PMMA was adapted as a substrate to fabricate holographic sensors. PMMA manufactured through casting was used since these substrates did not have birefringence and this was a requirement in Denisyuk reflection holography. Using an O_2 plasma cleaner, the substrates were treated under a vacuum of 1 torr for 3 min to render the PMMA surface hydrophilic. On these substrates, pHEMA matrices were synthesised by free radical polymerisation (Fig. 3.2a). $AgNO_3$ (1 M, 200 µl) dissolved in DI water was pipetted as an elongated blob onto a glass sheet. pHEMA matrix side down, the slide was placed on the elongated blob and the slide was incubated in situ for 3 min to allow diffusion of $AgNO_3$ into the pHEMA matrix (Fig. 3.2b). The sample was dried under a tepid air current for 10 s. The pHEMA matrix was immersed in a

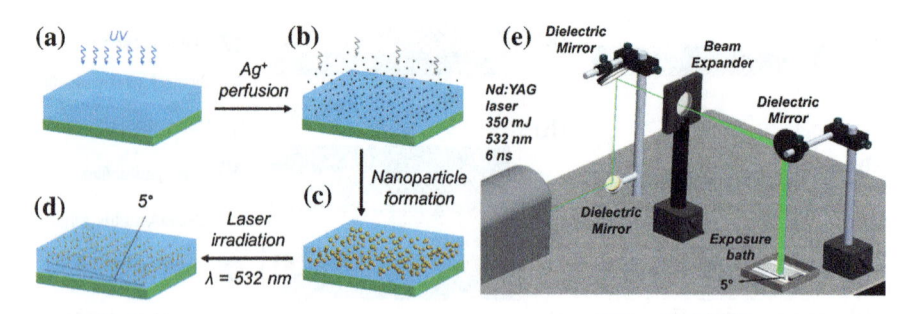

Fig. 3.2 Fabrication of holographic sensors by in situ size reduction. **a** HEMA monomers, ethylene dimethacrylate and methacrylic acid were copolymerised on an O_2-plasma-treated PMMA substrate. **b** Ag^+ ions were perfused into pHEMA matrix. **c** Using a photographic developer, Ag^+ ions were reduced to Ag^0 NPs in the pHEMA matrix. Reprinted with permission from Ref. [18] Copyright 2014 Wiley-VCH Verlag GmbH&Co. KGaA, Weinheim. **d–e** Patterning the hydrogel matrix in "Denisyuk" reflection mode using a Nd:YAG pulsed laser. Reproduced from Ref. [24] with permission from The Royal Society of Chemistry

photographic developer until no more darkening was seen (Fig. 3.2c). The sample was washed thoroughly with tap water, prior to immersion in acetic acid (5 % v/v) to neutralise the developer. A levelled Petri dish, with a silver mirror in the bottom surface, was filled with ascorbic acid (unbuffered, 2 % w/v, pH 2.66) and with the pHEMA side facing down, the sample was immersed in the bath at an inclination of 5. The pHEMA matrix remained in the Petri dish for 15 min to equilibrate. The sample was exposed to a single 6 ns pulse, Q-switch delay set to 258 μs by Nd: YAG pulsed laser ($\lambda = 532$ nm, 350 mJ) with a spot size of 2.5–3.5 cm (Fig. 3.2d). Figure 3.2e shows the laser writing setup in Denisyuk reflection mode. In order to produce holographic flakes, the slides were immersed in water-ethanol solution (50 %, v/v) for 1 h. After the flakes floated off the PMMA slide, they were stored in a glass Petri dish containing DI water to prevent sticking.

3.3   Characterisation of Holographic pH Sensors

Holographic pH sensors fabricated by in situ size reduction of Ag^0 NP were characterised. Optical and electron microscopes allowed investigating the size and the distribution of Ag^0 NPs throughout the polymer matrix. Angular-resolved measurements were taken to understand the sensor's diffractive properties. Other characterisation methods such as diffraction efficiency, index of refraction, and surface roughness and thickness measurements were correlated with the micrographs. Finally, limitations of this fabrication approach and its characterisation methods were pointed out.

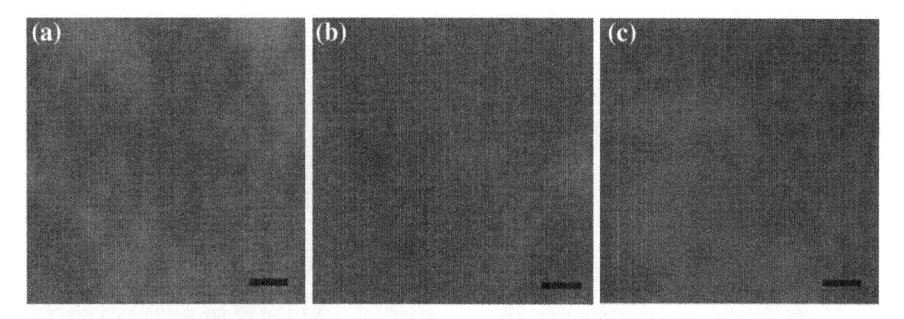

Fig. 3.3 Optical images of the surface gratings of pHEMA + Ag^0 NP system, showing three spots of different holograms. Scale bars = 10 μm

3.3.1 Microscopic Imaging of Holographic pH Sensors

Optical images of the surface topography of holographic sensors were taken using a Nikon OPTIPHOT-2 microscope in bright field mode. The surface gratings of the holographic pH sensors were analysed by studying the images of pHEMA + Ag^0 NP system (Fig. 3.3). Using the grating equation and measuring the position of the order from the images, the grating spacing was calculated as 3.01 μm. These gratings are transmission holograms caused by the interference of the object beam and internally reflected of light in Denisyuk reflection mode [25, 26].

Samples before (control) and after photochemical patterning were imaged. HELIOS 600 SEM/FIB was used for scanning electron microscopy (SEM). Squares were cut out of the matrices and stuck to a SEM Al stub using carbon tab/tape. An Au^0 film (~10 nm) was sputtered on the top surface to improve conductivity. A trench was prepared using a focused gallium ion beam milling. The cleaning was achieved through low current milling. Figure 3.4a, b illustrates cross sectional images of the pHEMA-co-MAA + Ag^0 NP system before laser-induced photo-chemical patterning (control samples), and Fig. 3.4c, d shows the system after patterning. The Ag^0 NPs are situated up to ~5–6 μm below the polymer matrix's surface (Fig. 3.4a). However, SEM was not suitable for polymer imaging since the electron beam generated heat as it hit the specimen, and this damaged the polymer and generated bubbles (Fig. 3.4c, d). Additionally, the melting temperature (T_m) is size dependent. Ag^0 NPs with diameters of 5, 11, 14 and 20 nm melt at 480, 550, 580 and 650 °C, respectively [27]. Since laser-induced photochemical patterning attenuates the size of the Ag^0 NPs [28], Fig. 3.4c–d have smaller Ag^0 NPs, hence subject to melting as compared to the polymer matrices before patterning.

Environmental Scanning Electron Microscopy (ESEM) was used to image the cross section of the pHEMA-co-MAA + Ag^0 NP system. Holographic sensors on PMMA were cross-sectioned using a microtome. Images were obtained with a Philips XL30 FEG ESEM, acceleration voltage = 6 kV with a working distance (WD) of 13.5 mm. An advantage of ESEM is that images can be obtained at room temperature (24 °C). Using ESEM, dynamic swelling experiments were performed in wet mode.

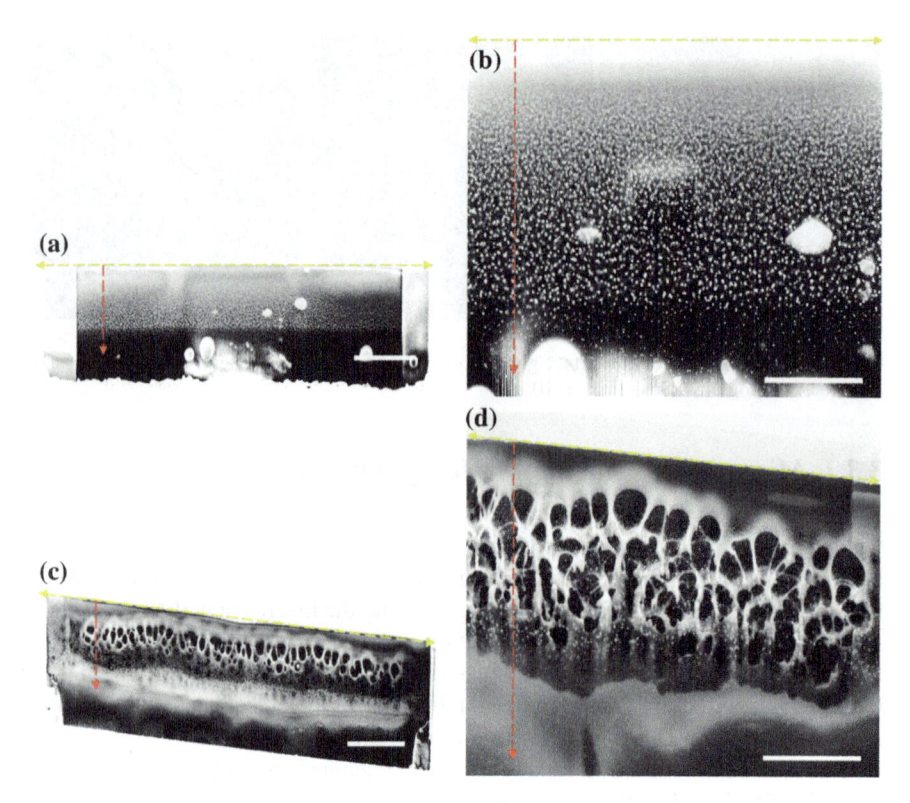

Fig. 3.4 Cross sectional images of the pHEMA + Ag^0 NP system, **a–b** before and **c–d** after laser-induced photochemical patterning (control samples). The *yellow dashed lines* represent the surface of the pHEMA matrix. The *red dashed lines* show the approximate depth of Ag^0 NP penetration into the pHEMA matrix. The contrast and the brightness were increased to distinguish the Ag^0 NPs from the pHEMA matrix. Scale bars in **a–c** are 5 μm and in **b–d** are 2 μm

The Peltier heating/cooling stage allows imaging ±20 °C room temperature. Figure 3.5a–d shows the cross sectional images of the pHEMA + Ag^0 NP system. The yellow, red and blue dashed lines represent (i) the surface of the pHEMA matrix, (ii) the approximate depth of Ag^0 NP penetration into the pHEMA matrix, and (iii) the pHEMA-PMMA interface. As the pressure was increased from 1.6 to 6.0 torr at 2–6 °C, the polymer absorbs water and expands from 8.9 to 10.1 μm normal to its underlying substrate, a change of 13.5 % (Fig. 3.5e). At 6.0 torr, the relative humidity is 100 %.

Transmission Electron Microscopy (TEM) images were also taken. Hydrogels were released from their substrates and transferred to dry ethanol. This was replaced with 2 changes of acetonitrile (CH$_3$CN) for 10 min each. They were transferred from dry ethanol into a mixture of acetonitrile and Araldite epoxy resin (1:1, v/v). The acetonitrile was allowed to evaporate overnight. The hydrogels were transferred through 3 changes of Araldite for 2 h each and the resin was cured at 60 °C for 48 h.

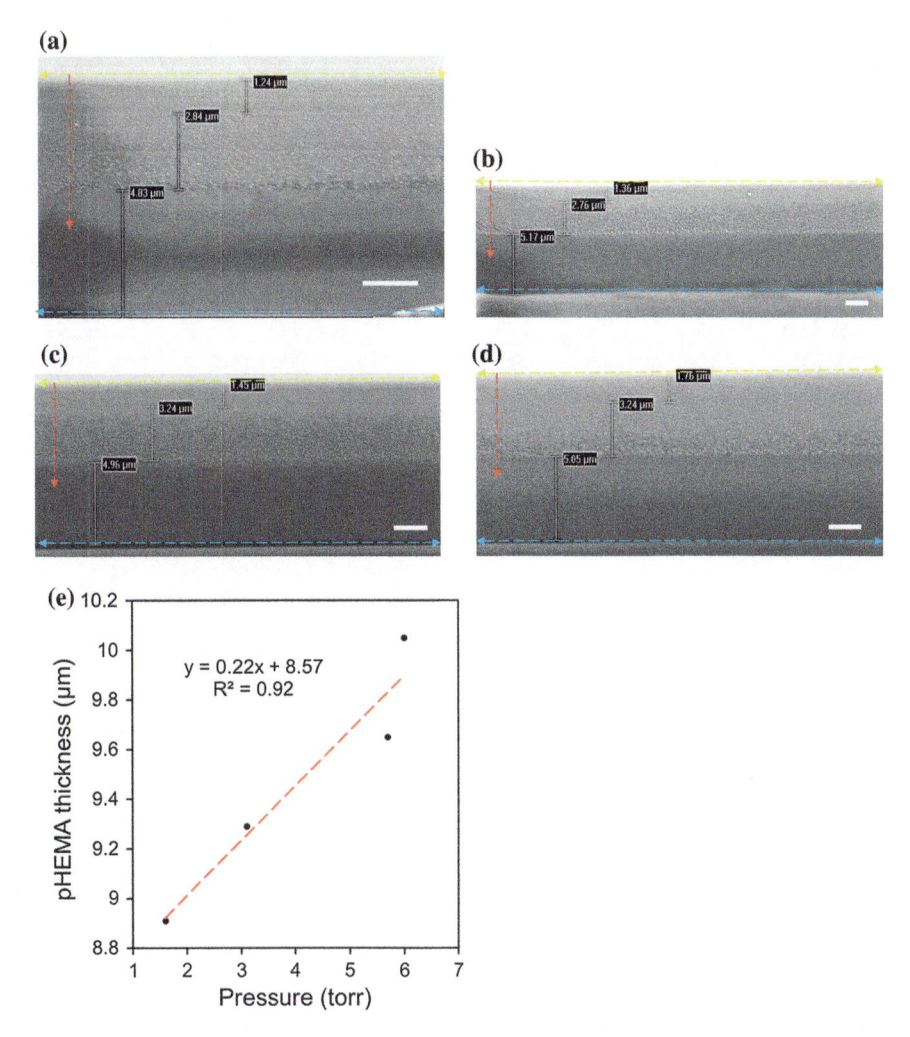

Fig. 3.5 Environmental scanning electron microscopy images of the holographic matrix. Under **a** 1.6, **b** 3.0, **c** 5.7 and **d** 6 torr. The *yellow, red and blue dashed lines* represent (i) the surface of the pHEMA matrix, (ii) the approximate depth of Ag^0 NP penetration into the pHEMA matrix, and (iii) the pHEMA-PMMA interface. The contrast and the brightness were increased to distinguish the Ag^0 NPs from the pHEMA matrix. **e** The change in the thickness of pHEMA + Ag^0 NP matrix due to an increase in pressure and water absorption

Vertical sections through the hydrogel were cut with a diamond knife (Deyemond, Germany) using a Leica Ultracut UCT (Leica, Vienna). They were mounted on 400 mesh copper grids and viewed in a FEI Tecnai G2 (Oregon, USA) operated at 120 kV. Images were recorded with an AMT 60B camera running Deben software. Figure 3.6 illustrates cross-sectional images of the pHEMA + Ag^0 NP system before laser-induced photochemical patterning (control samples), and Fig. 3.7 shows the

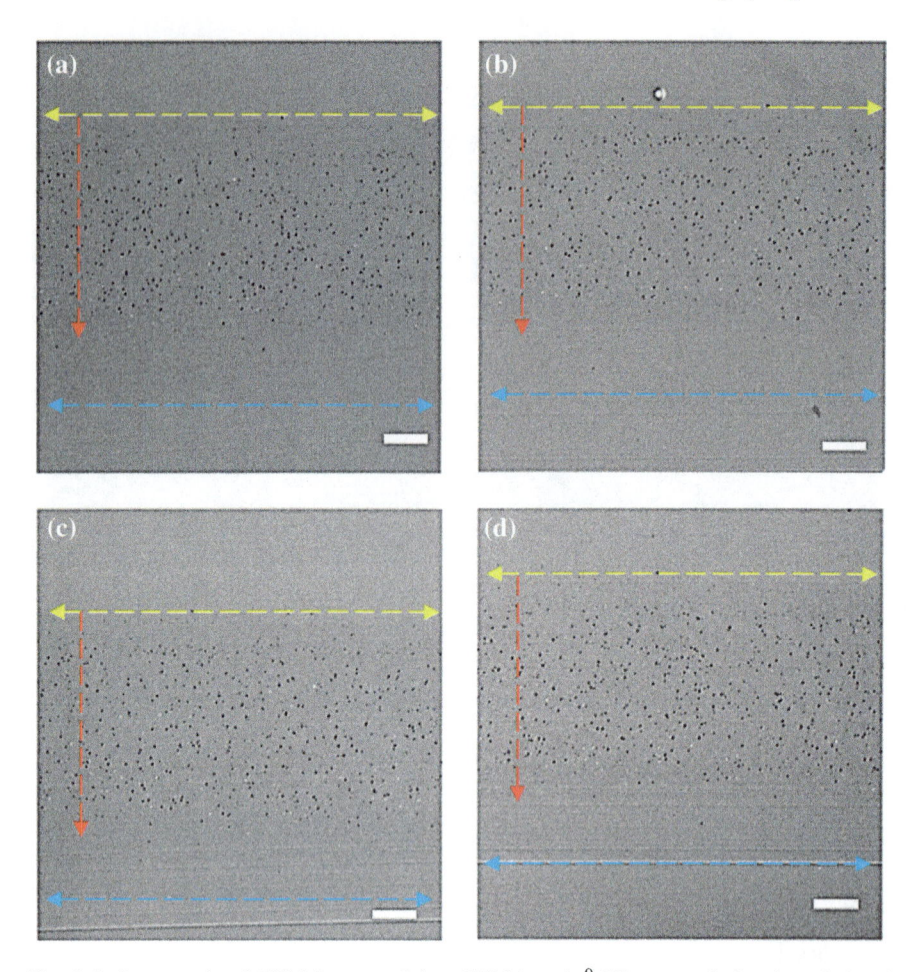

Fig. 3.6 Cross sectional TEM images of the pHEMA + Ag^0 NP system before laser-induced photochemical patterning (control samples). **a–d** Cross sections of different samples. The *yellow, red and blue dashed lines* represent (i) the surface of the pHEMA matrix, (ii) the approximate depth of Ag^0 NP penetration into the pHEMA matrix, and (iii) the pHEMA-PMMA interface. The contrast and the brightness were increased to distinguish the Ag^0 NPs from the pHEMA matrix. Scale bars = 2 μm

system after patterning. The yellow, red and blue dashed lines represent (i) the surface of the pHEMA matrix, (ii) the approximate depth of Ag^0 NP penetration into the pHEMA matrix, and (iii) the pHEMA-PMMA interface. The Ag^0 NPs can penetrate ~5–6 μm into the polymer matrix. TEM generates less heat than SEM, while allowing high-resolution imaging of Ag^0 NPs.

The TEM images were analysed to measure the NP size distribution before and after laser-induced patterning. Standard deviation values for the diameter of the Ag^0 NPs before (Fig. 3.8a, b) and after (Fig. 3.8c, d laser patterning were 34 ± 23 nm (n = 96) and 18 ± 12 nm (n = 1162), respectively. Figure 3.8e illustrates the

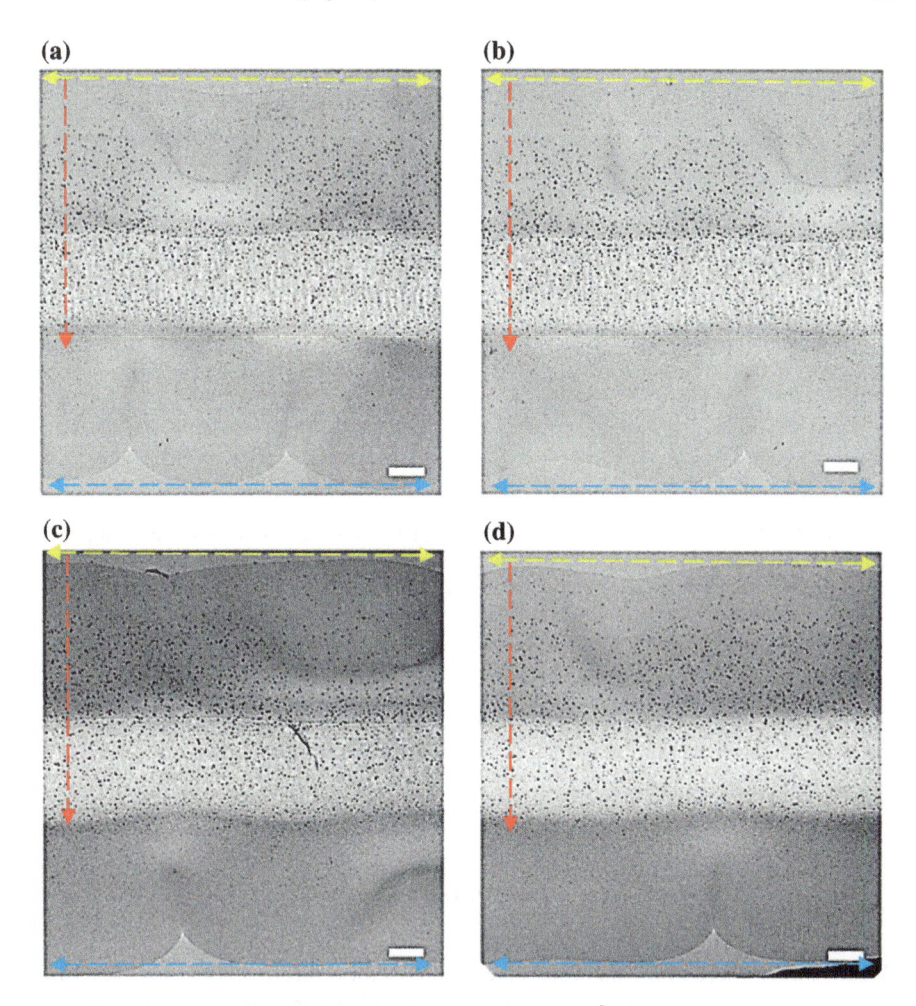

Fig. 3.7 Cross sectional TEM images of the pHEMA + Ag^0 NP system after laser-induced photochemical patterning. **a–d** Series of cross sections showing Ag^0 NP distribution. The *yellow, red and blue dashed lines* represent (i) the surface of the pHEMA matrix, (ii) the approximate depth of Ag^0 NP penetration into the pHEMA matrix, and (iii) the pHEMA-PMMA interface. The contrast and the brightness were increased to distinguish the Ag^0 NPs from the pHEMA matrix. Scale bars = 2 μm

distribution of these Ag^0 NPs across the cross section of the pHEMA matrix ($\sim$5–10 μm below the polymer surface). The brightness difference between the non-patterned and patterned regions might be due thermal or physical degradation of the polymer. The NP size distribution measurements might also be influenced by the position in the plane. The laser wavelength, power, pulse duration, NP size and surface plasmon resonance may affect the mechanism of laser-induced photochemical patterning and subsequent absorption by the Ag^0 NPs [29–31].

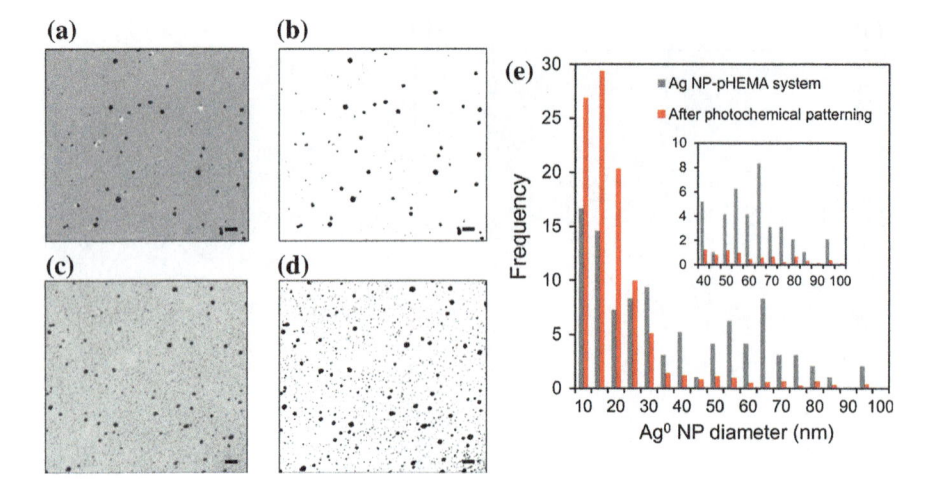

Fig. 3.8 TEM images of the cross sections of the pHEMA + Ag^0 NP system representing non-patterned and patterned regions at ~ 5–10 μm below the polymer surface. **a** Photographic image image, **b** Threshold before laser-induced patterning, **c** Photographic image, **d** Threshold after laser-induced patterning. Scale bars 200 nm. **e** Size distributions of Ag^0 NPs before and after laser-induced photochemical patterning in pHEMA matrix

The interaction between the laser light and the NPs in situ may also influence particle diffusion, oxidation, structure and distribution [32–34]. The outlined factors above contribute to organising the NPs in the polymer matrix.

3.3.2 Effective Index of Refraction Measurements

While direct measurements were taken by an Abbé refractometer with an LED, indirect measurements were obtained using a plano-convex lens. The pHEMA matrices were lifted off from their substrates, lubricated with an index matching fluid, and placed on the reading plate of the refractometer. Indirect measurements of the index of refraction were obtained by the consecutive focus point distance measurements of a plano-convex lens immersed in various solutions and pHEMA layer. The focal points at air (n = 1.00), water (n = 1.33), ethanol (n = 1.36) and decane (n = 1.41) were measured by placing the lens, convex side down, in a glass Petri dish, and submerging the lens into these solvents. A monomer mixture consisting of HEMA (91.5 mol%), EDMA (2.5 mol%), MAA (6 mol%) solution was polymerised with a lens in the Petri dish, followed by the measurement of focal length of water saturated-hydrogel system. Finally, the index of refraction of water-hydrogel system was extrapolated based on the reference focal length measurements of air, water, ethanol and decane. The effective index of refraction of the non-patterned hydrogel (pHEMA, Ag^0 NPs) was 1.46 ± 0.01, and it decreased to 1.43 ± 0.01 after laser-induced

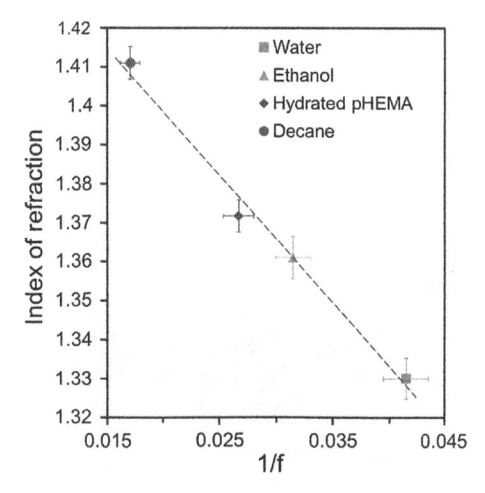

Fig. 3.9 Indirect estimation of effective index of refraction of pHEMA matrix using a plano-convex lens

photochemical patterning. Indirect measurement of the index of refraction of water-pHEMA system was 1.37 ± 0.03, obtained using a plano-convex lens (Fig. 3.9).

3.3.3 Angular-Resolved Measurements

In order to characterise the spectral response of the underlying photonic structure, the holographic sensor was illuminated at normal incidence and 45° from the normal with a supercontinuum white light laser (Fig. 3.10a, b). The laser operates in the MHz repetition rate range with picosecond pulses. Its spectral range spans from ~400 nm to visible to near-infrared. The backscatter was recorded at different angles (0–180°) by a spectrophotometer positioned onto a goniometer, which was varied incrementally at small angles (~1°). Figure 3.10c–e illustrates the spectra recorded for different collection angles and the intensity of the peak (in violet-to-red colour scale). The grating structure diffracts light analogous to a blazed (saw tooth) grating, which corresponds to the transmission hologram. However, the diffraction is two orders of magnitude more efficient at ~530 nm compared with the other diffracted wavelengths due to the underlying multilayer structure parallel to the grazing surface. The red regions in Fig. 3.10c–e show the angle positions, where the diffracted light intensity is the highest due to the underlying multilayer structure. However, the continuous grating green and cyan few angles away from the multilayer grating indicate the transmission hologram. The red regions in the bottom part of the figure show a saturated signal since at 0–5° the receiver of the spectrophotometer is saturated by the laser light. The combination of these two structures effectively provides an intense colour-selective backscatter at angles away from the sample normal. The zero order of diffraction (0°) is saturated to highlight the presence of the first and second orders of diffraction. The intensity is in a logarithmic scale.

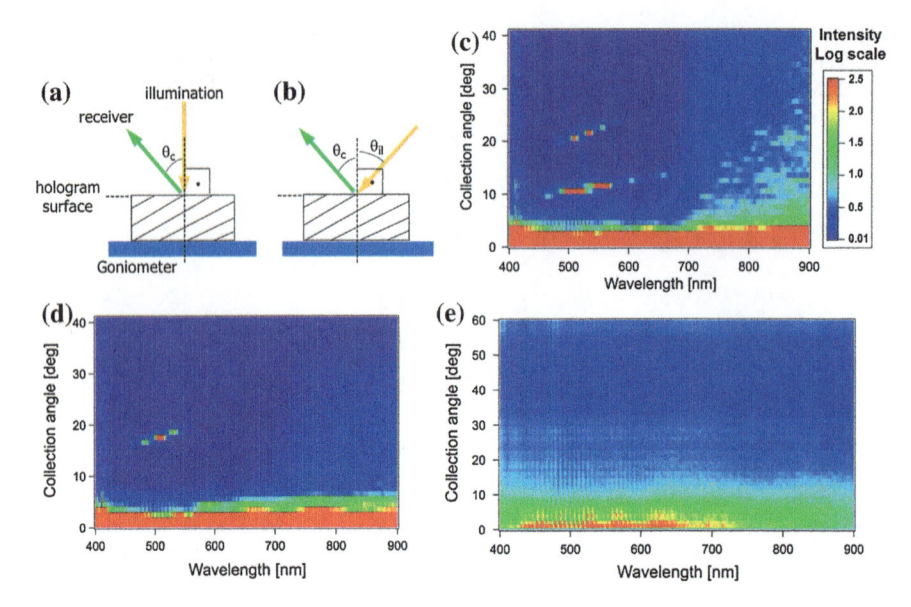

Fig. 3.10 The optical setup for angular-resolved measurements of the diffracted light from the photonic structure. Supercontinuum white light illumination, **a** normal, **b** angle with respect to the hologram surface. Diffraction measurements at, **c** 0°, **d** 45°, **e** specular reflection measurements from 0–60° angle of incidence

3.3.4 Diffraction Efficiency Measurements

The power output of the pulsed Nd:YAG laser (350 mJ specified by the manufacturer) was measured using a powermeter (Fieldmaster, 30 V). The beam exiting the laser was directed by dichroic mirrors and passed through a beam expander and a collimating lens before entering the powermeter receiver elevated 8 cm above the optical table. The measurements were made by producing single pulses at Q-switch values from 450 to 258 μS. Holograms comprising pHEMA + Ag^0 NP were prepared using different laser pulse energies. The samples were exposed to subsequent (1–20) Nd:YAG laser (6 ns, 532 nm) pulses. Diffraction efficiency measurements were conducted to find out the effect of laser light in the organisation of the Ag^0 NPs. These measurements were performed at the same laser wavelength, and the diffracted and refracted lights were measured using a powermeter. First, the laser power output of the laser was recorded. As the quality factor decreased, the power output of the Nd:YAG pulsed laser increased exponentially (Fig. 3.11a). The results show that the highest diffraction efficiency was obtained after three laser pulses (Fig. 3.11b). This may be attributed to the ability to gradually organise Ag^0 NPs within the polymer matrix. However, multiple high-energy laser pulses may damage the polymer matrix.

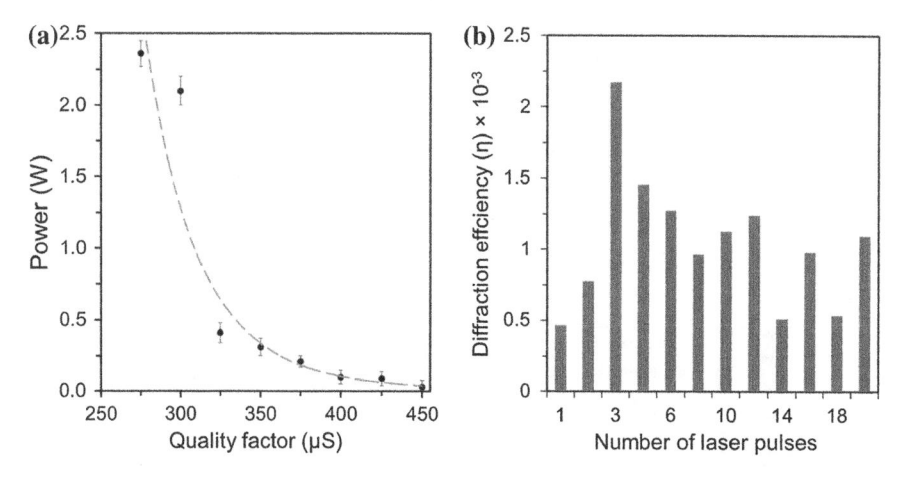

Fig. 3.11 Laser power output and diffraction efficiency measurements. **a** The power output of the laser as a function of quality factor. As the quality factor decreases, the power output of the laser increases. **b** Diffraction efficiency of the pHEMA + Ag^0 NP system formed at different laser light pulses

3.3.5 Polymer Thickness and Roughness Measurements

The profilometer is an advanced thin film step height measurement device that has a step accuracy below 100 Å to profile surface topography and waviness, as well as measuring surface roughness in the sub-nm range. The device electrochemically measures the sample by moving it beneath a diamond-tipped stylus. A stage moves the sample based on a given scan length, speed and stylus force. The stylus is mechanically attached to the core of a linear variable differential transformer (LVDT). When the stage moves the substrate and its attached polymer matrix, the stylus rides over the surface of the matrix. Variations in the surface roughness move the stylus vertically, which probes an electrical signal based on the degree of the corresponding stylus movement in LVDT. LVDT scales an AC reference signal proportional to the position change and converts it from analog to digital format. The profilometer was set to a mode to measure the desired range. To confirm the accuracy of the measurement, the probe was moved at different speeds. The measurements were carried out with matrices consisting of (i) pHEMA, (ii) pHEMA impregnated with Ag^0 NPs and (iii) pHEMA impregnated with Ag^0 NPs after laser-induced photochemical patterning in Denisyuk reflection mode. Figure 3.12a shows the measurement locations. Batch to batch variation was also determined. The pure pHEMA thickness was 7.82 ± 3.12 µm (n = 18) (Fig. 3.12b). This decreased to 6.26 ± 1.88 µm (n = 18) after the system was impregnated with Ag^0 NPs. After photochemically patterning the system, the thickness was 5.20 ± 1.40 µm (n = 6). The decrease in the polymer thickness can be attributed to a decrease in Ag^0 NP size. This may be due to accommodation of attenuated particles in smaller pores. The surface roughness of the pure pHEMA was measured to be 43.19 ± 13.80 nm

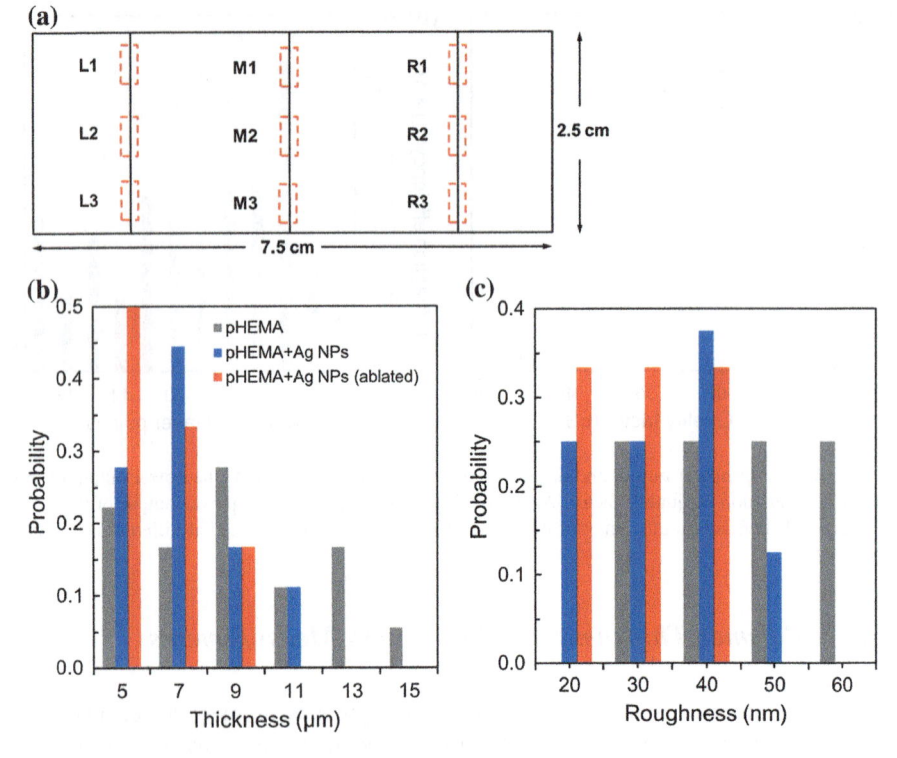

Fig. 3.12 Characterisation of pHEMA and Ag⁰ NPs system before and after photochemical patterning. **a** The measurement zones of polymer matrices across a microscope slide in profilometry, **b** Thickness and **c** roughness measurements

(n = 9) and it decreased to 29.84 ± 9.97 nm (n = 8) after the system was impregnated with Ag^0 NPs (Fig. 3.12c). After photochemically patterning the surface roughness was 21.47 ± 10.72 nm (n = 3). A plausible explanation for the increase in the roughness after photochemical patterning is that laser writing might damage the residual Ag^0 layer left on the surface of the pHEMA matrix during Ag^0 NP formation.

3.4 Optical Readouts

3.4.1 Holographic pH Sensors Fabricated Through Silver Halide Chemistry

The holographic sensor was functionalised with methacrylic acid (MAA) that reversibly bound H^+ ions. The deprotonation, and hence ionisation, of carboxyl groups, increased the Donnan osmotic pressure within the pHEMA matrix.

Consequently, this resulted in water uptake, swelled the hydrogel and increased the spacing of the Ag^0 NPs, primarily in the vertical (and to a much lesser extent in the lateral) direction, which shifted the Bragg peak to longer wavelengths. Holographic sensors (0.5×2.5 cm) were submerged in a plastic cuvette (reservoir) for pH measurements. After a reading was taken, the reservoir solution was removed and the cuvette was flushed consecutively three times for each reading point. A spectrophotometer in reflection mode was used with a bifurcated cable to capture the corresponding images. Figure 3.13a shows the measurements of Bragg peaks of the holographic pH sensor fabricated using silver halide chemistry. As the pH was increased, the Bragg peak shifted to longer wavelengths by ~ 280 nm (Fig. 3.13b). An increase in lattice spacing consequently reduced the contrast of effective index of refraction, hence the efficiency of the multilayer structure decreased. A modified Henderson-Hasselbalch equation was adapted to determine the apparent pK_a values [35]:

$$\lambda_{\text{shift}} = \frac{\Delta \lambda}{\left(10^{(pK_a - pH)} + 1\right)} \tag{3.1}$$

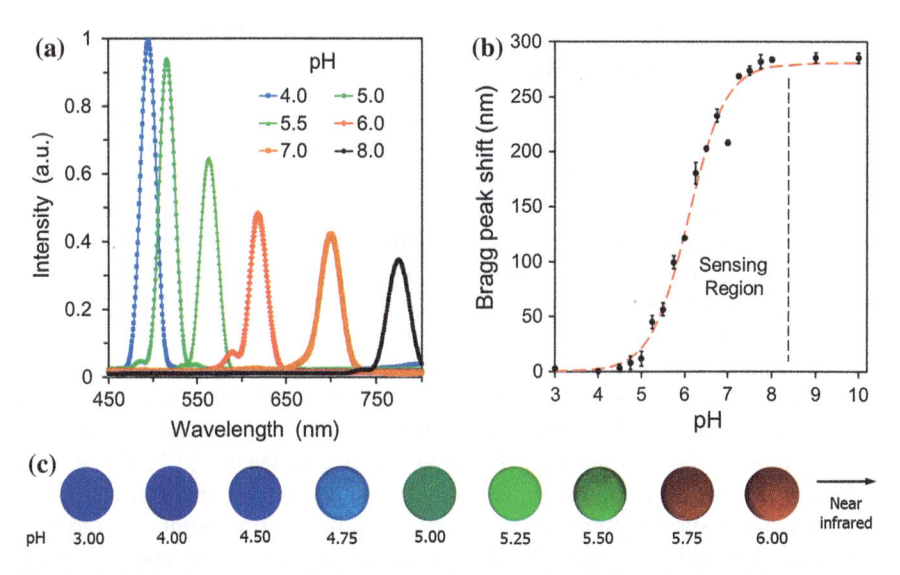

Fig. 3.13 Diffraction spectra of holographic pH sensors (6 % MAA) fabricated through silver halide chemistry. **a** Visible-near-infrared diffraction spectra of a holographic sensor swollen by different pH solutions using phosphate buffers at 24 °C, **b** The Bragg peak shifts over three trials, **c** Colorimetric readouts of the holographic sensor at various pH values. Na_2HPO_4-citric acid (pH 3.00–8.00), Na_2HPO_4-HCl (pH 9.00), Na_2HPO_4-NaOH (pH 10.00) were mixed to obtain buffers (150 mM) at desired pH values. Standard errors were calculated from the three subsequent trials for each data point. Reproduced from Ref. [17] with permission from The Royal Society of Chemistry

where λ_{shift} is the Bragg peak shift, $\Delta\lambda$ is the difference between the observed maximum and minimum Bragg peak positions, and pK_a is the acid dissociation constant. The apparent pK_a value was calculated as 6.08 using the Henderson-Hasselbalch equation. Figure 3.13c illustrates the readouts from a digital camera showing different colours diffracted at pH values from 3.0 to 6.0. The sensor operated within the visible spectrum as well in the near infrared. The spectrophotometer had a resolution of 0.5 nm wavelength shift, which corresponded to a minimum lattice swelling distance of 0.18 nm, obeying Bragg's law. A ~ 10 μm thick hydrogel, which can theoretically accommodate ~ 55 fringes (or stacks), needs to swell a minimum of ~ 9.7 nm in order to cause a resolvable spectral shift. The overall error in Bragg peak measurement stemmed from multiple sources [36]. Negligible instrumental errors included the pH meter resolution (~ 0.05 pH) and spectrophotometer resolution (0.5 nm) (Mettler Toledo, User Manual). Other errors might be due to pipetting, uneven polymerisation, heterogeneity of functional groups or the crosslinker in the monomer solution. The error of the overall system could be mitigated by building a calibration curve with smaller pH increments, increasing the number of replicates, and incorporating the effect of temperature and humidity in the calibration of the holographic sensor.

3.4.2 Holographic pH Sensors Fabricated Through in Situ Size Reduction of Ag^0 NPs

As the concentration of H^+ ions decreased from pH 4.00 to 8.00 under physiological (150 mM) ionic strengths, the Bragg peak originating from the multilayer structure systematically shifted from 495 to 815 nm (Fig. 3.14a, b). The shift in the Bragg peak was visible to the eye throughout most of the measurement (Fig. 3.14c). As the hologram expanded, the diffraction efficiency decreased (the intensity of the peaks). This trend could be attributed to the decrease in the density of Ag^0 NPs present in the periodic regions of the hologram, which reduced the effective refractive index contrast between these regions and the medium. Another contributing factor was that the scattering strength of each Ag^0 NP increased at Mie plasmon resonances in the blue/green region, so the total amount of scattering decreased as the Bragg resonance shifted to longer wavelengths [37]. The sensing process was reversible and the same sensor might be used for multiple analyses.

Consecutive swelling/shrinking steps were reproducible to within ±3 nm over 20 successive buffer changes. No hysteresis was detected. For example, a pH change of 0.50 units from pH 5.50 to 6.00 took 50 ± 10 s to reach 90 % of the maximum signal, with a variation of Bragg peak shift of ±1 nm over three trials. The pH-sensing range and the sensitivity of the sensor can be controlled through the variation of the ionisable co-monomer in the polymer matrix and its concentration, respectively. For instance, to achieve sensitivity at acidic and alkaline pH, the polymer matrix can be functionalised with MAA and 2-(dimethylamino)ethyl

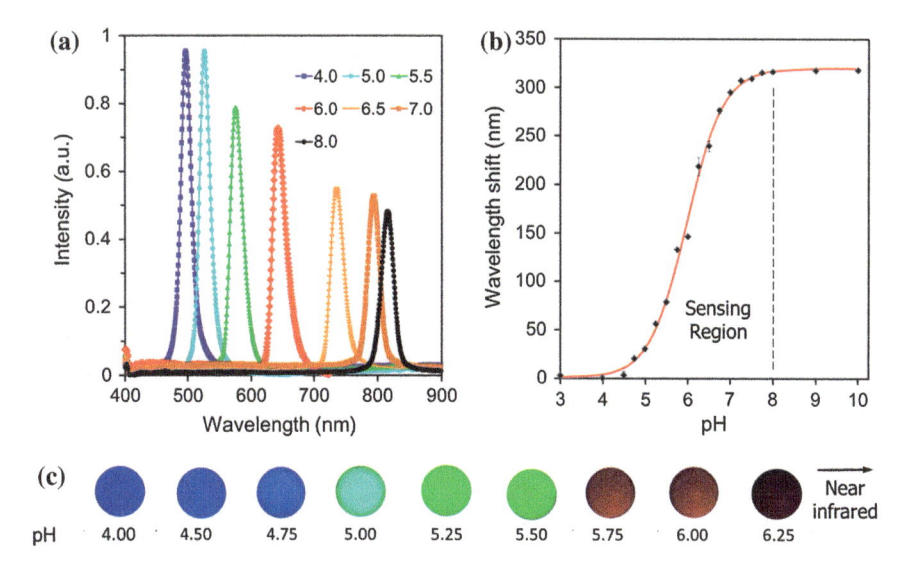

Fig. 3.14 Tuning the holographic pH sensor (6 % MAA) fabricated through in situ size reduction of Ag^0 NPs. **a** Visible-near-infrared diffraction spectra of the holographic sensor swollen by different pH values using phosphate buffers (150 mM) at 24 °C. The smallest Bragg peak was at 495 nm and the largest was at 815 nm—a change of 165 %. **b** Sensor response over three trials. The apparent pK_a value was calculated as 5.98 using the Henderson-Hasselbalch equation. **c** Photographs of the sensor immersed in phosphate buffers of pH 4.00–6.25. The images were taken under white light illumination. Reprinted with permission from Ref. [18] Copyright 2014 Wiley-VCH Verlag GmbH&Co. KGaA, Weinheim

methacrylate (pK_a = 8.40) [20]. In order to extend the range, a mixture of ionisable monomers may be co-polymerised. Such hydrogels can also be functionalised to be highly selective to a range of stimuli [24, 38–45]. Figure 3.15 illustrates the summary of holographic pH sensors fabricated on different substrates. The photochemical patterning through in situ size reduction of Ag^0 NPs was comparable to the previously developed silver halide chemistry-based holographic pH sensors. However, the maximum achievable Bragg peak shifts differed in three cases (Fig. 3.15). This might be attributed to the error in controlling MAA concentration and degree of crosslinking in each formulation.

3.4.3 Interference Due to Metal Ions

Biological samples such as urine and blood contain a mixture of ions and other potential interferents. The effect of these interferents on the holographic pH sensors was evaluated. Control experiments were conducted to assess the sensitivity of pHEMA (no MAA) to Na^+, K^+, Mg^{2+} and Ca^{2+} ions. Na^+ ions are the most

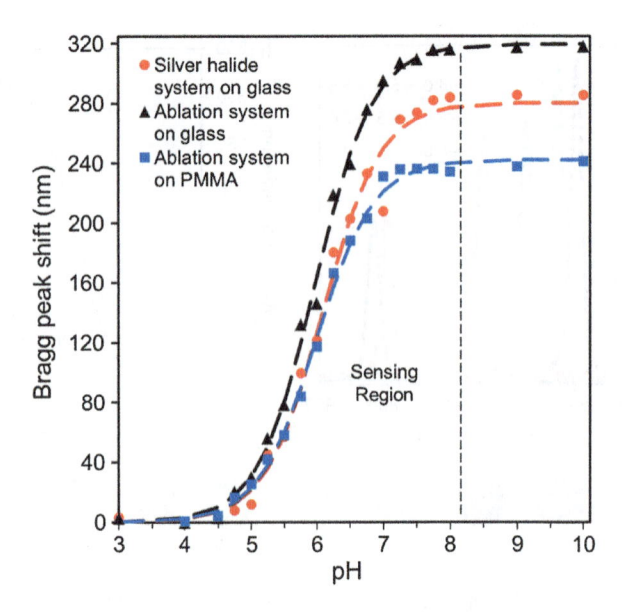

Fig. 3.15 Bragg peak shifts of holographic pH sensors on glass and PMMA substrates at 24 °C. Apparent pK_a values were calculated as 6.08, 5.97, 5.98 for silver halide system on glass, in situ size reduction of Ag^0 NPs (ablation system) on glass and PMMA, respectively

abundant metal ion in biological fluids followed by K^+ ions. An increase in Na^+ ion concentration from 0 to 200 mM caused a blue Bragg shift of 14 nm (Fig. 3.16a). Similarly, an increase in K^+ ion concentration from 0 to 100 mM caused a blue Bragg shift of 9 nm (Fig. 3.16a). Changes in divalent metal ions Ca^{2+} and Mg^{2+} from 0.0–8.0 mM and 0.0–10.0 mM resulted in blue Bragg peak shifts of 2.4 and 2.9 nm, respectively (Fig. 3.16b).

3.4.4 Ionic Strength Interference in pH Measurements

The effects of the pH and the ionic strength (Na^+ ions) were investigated for the pHEMA matrix with residual carboxylic acid groups. HEMA monomers may polymerise in the absence of a stabiliser and small amounts of diethylene glycol monomethacrylate, di(ethylene glycol)dimethacrylate, methacrylic acid, which are included in its manufacturing to stabilise this monomer. Variation in Bragg peak shifts up to ~ 6 nm was measured for each ionic strength (25–150 mM) value as the pH was increased 4.5–7.0 (Fig. 3.17). These effects were caused both Donnan osmotic pressure and pH changes. However, the pH effect can be minimised by passing the monomer HEMA through an Al_2O_3 column.

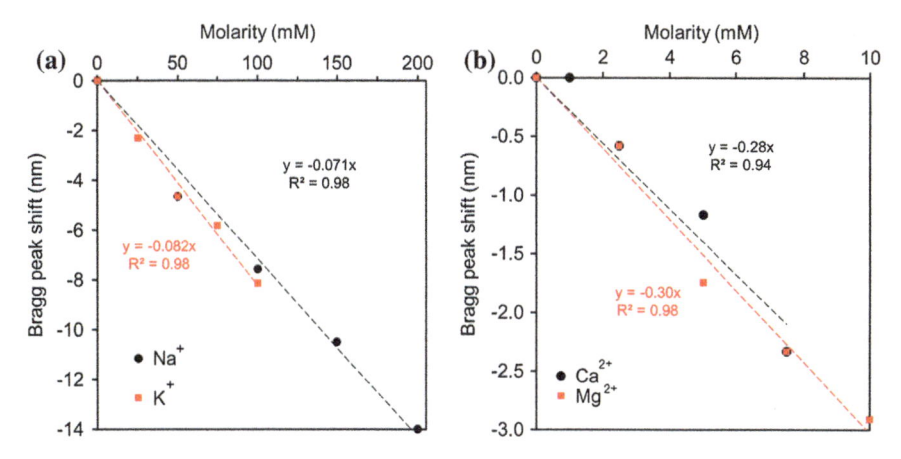

Fig. 3.16 Interference due to monovalent and divalent metal ions in holographic pH sensor response at 24 °C. **a** Na^+ and K^+, **b** Ca^{2+} and Mg^{2+} ions

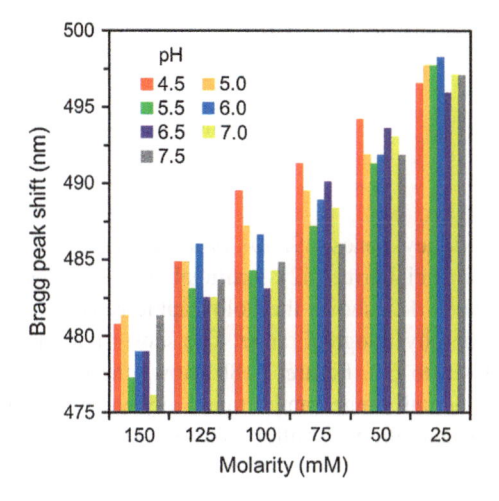

Fig. 3.17 Bragg peak shifts due to pH and Na^+ ion concentration as the molarity was changed in pHEMA matrices with residual MAA remaining from its manufacturing process

3.4.5 Sensing pH in Artificial Urine

To demonstrate the utility of the holographic pH sensor, a clinical application to analyse acid-base balance in artificial urine samples is described. Urine pH in a healthy individual is in the range of 5.0–9.0 [46]. Analysis of urine pH has diagnostic utility in, for example, the evaluation of urinary tract infections (UTIs) [47] and renal tubular acidosis [48]. Modification of the urine pH allows the treatment

Table 3.1 Composition of artificial urine

Component	Quantity (g)	Concentration (mmol l^{-1})
Peptone L37	1	N/A
Yeast extract	0.005	N/A
Lactic acid	0.1	1.1
Citric acid	0.4	2
Sodium bicarbonate	2.1	25
Urea	10	170
Uric acid	0.07	0.4
Creatinine	0.8	7
Calcium chloride. $2H_2O$	0.37	2.5
Sodium chloride	5.2	90
Iron II sulphate. $7H_2O$	0.0012	0.005
Magnesium sulphate. $7H_2O$	0.49	2
Sodium sulphate. $10H_2O$	3.2	10
Potassium dihydrogen phosphate	0.95	7
Dipotassium hydrogen phosphate	1.2	7
Ammonium chloride	1.3	25
Distilled water	To 1L	N/A
HCl	To specific pH	1
NaOH	To specific pH	1

for certain types of kidney stones [49, 50], or toxic ingestions [51]. Table 3.1 shows the composition of artificial urine [52]. Artificial urine samples were measured from pH 4.5 to 7.0. Figure 3.18a shows the holographic sensor's Bragg peak shift as a function of pH in the physiological range. The sensor had high sensitivity (48 nm/ pH unit) from pH 4.6 to 6.6. Figure 3.18b shows the corresponding colorimetric response. This wavelength range can be tuned anywhere from UV to near infrared by changing the concentration of buffer solution used during laser exposure.

3.4.6 Paper-Based Holographic pH Sensors

Holographic flakes can be integrated with paper-based strips to enable wicking aqueous samples. However, the white background of the paper was not preferable since it reflected the incident light, which interfered with the diffracted light. Hence, a number of strategies was developed to reduce the background noise by dyeing the paper to black. Placing the hologram on a dark surface improved the signal-to-noise ratio, which increased the diffraction efficiency relative to the background. Paper towel-water emulsion (1:3, w/v) was ground using a blender, and mixed with poly (vinyl alcohol) (PVA) and Fe_3O_4 (500:1:4, v/w/w). After mixing, the pulp was

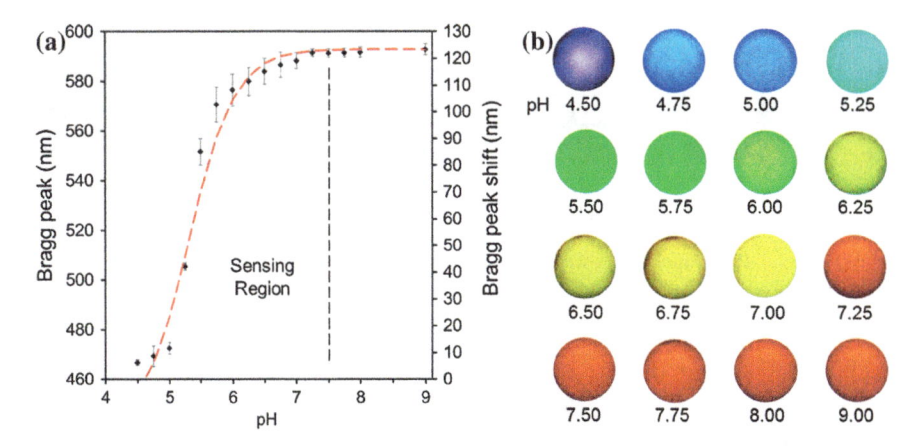

Fig. 3.18 Sensing pH changes in artificial urine solutions using the holographic sensor fabricated through in situ size reduction of Ag^0 NPs. **a** Bragg peak shift of the sensor as a function of different pH values of artificial urine samples over the physiological range. The apparent pK_a value was calculated as 5.20 using the Henderson-Hasselbalch equation. Standard error bars represent three independent samples. **b** Photographs of the holographic pH sensor readouts to artificial urine samples (pH 4.5–9.0). The images were taken under white light illumination. Reprinted with permission from Ref. [18] Copyright 2014 Wiley-VCH Verlag GmbH&Co. KGaA, Weinheim

poured on a sieve constrained by a deckle and subsequently dried. The resulting paper sheet was cut to 0.5 × 2 cm to produce paper strips. pHEMA holographic matrices were removed from the substrate, cut into flakes, and were assembled on iron oxide (Fe_3O_4) impregnated paper strips. Strips with pHEMA showed no deterioration such as cracking and breaking (Fig. 3.19a), but flakes with pure pHEMA-co-MAA did not attach to the surface of the paper (Fig. 3.19b, c). However, acetic acid (1 %, v/v) facilitated the attachment when the flakes were in contracted form. This prevented cracking of the pHEMA flakes when they were dried on the paper surface (Fig. 3.19d).

Chromatography and filter papers, and nitrocellulose were also dyed. A dye solution consisting of Procion MX-K and Na_2CO_3 in DI water (1:1:50, w/w/v) at 60 °C was prepared. Whatman® 1 chromatography paper, and Whatman® filter papers (type 5, 2, 597, 4) were immersed in the dye solution for 8 h at 24 °C. Dyed chromatography/filter papers were rinsed with DI water until excessive dye diffused out. The same process was repeated for Remazol Reactive Black 5, but the system was incubated at 50 °C. Two controls performed in this experiment included the effect of adding (i) Na_2CO_3, and (ii) urea (CH_4N_2O)—dilution of Procion Black MX-K (1:1, w/w). Sodium carbonate was essential in fixing Procion Black MX-K to paper (Fig. 3.20a). Remazol Reactive Black 5 needed to be rinsed with water (>1L) to remove the excessive dye (Fig. 3.20b). Urea, which normally acts as a humectant and keeps the fibres damp to allow a greater reaction time for Procion Black MX-K, was not necessary in dyeing (Fig. 3.20c, d) [53].

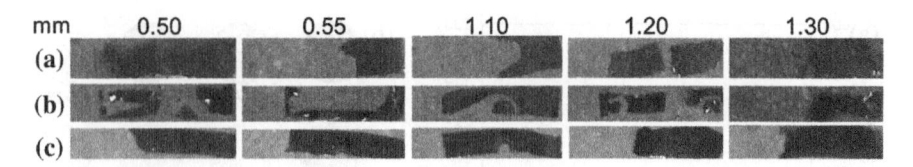

Fig. 3.19 Assembly of holographic flakes on paper-Fe$_3$O$_4$ composites. **a** pHEMA matrices, **b** pHEMA-co-MAA matrices, **c** pHEMA-co-MAA assembled to paper-Fe$_3$O$_4$ composite in acetic acid (1 %, v/v). Paper thicknesses were 0.50, 0.55, 1.10, 1.20 and 1.30 mm. Images were taken 6 h after assembly

Fig. 3.20 Dyeing cellulose-based materials with Procion Black MX-K and Remazol Reactive Black 5 at 24 °C. **a** Procion Black MX-K without sodium carbonate, **b** Remazol Reactive Black 5 with sodium carbonate before rinsing. **c** Procion Black MX-K with sodium carbonate and urea **d** Procion Black MX-K with sodium carbonate only

The composition of Procion MX-K was determined using thin-layer chromatography. Thin layer chromatography (TLC) sheets precoated with Al were cut to 1 × 5 cm. The plates were marked with a line drawn 1 cm from the bottom of the plate. Procion MX-K was diluted to 1 % (w/v) in DI water. A glass spotter (i.e. capillary tube with ∅ 1 mm) was used to deposit the dye in 1–2 mm spots on the plate. The plates were placed in a developing chamber saturated with the following eluents: (i) Hexane, (ii) ethanol:water (1:1, v/v), (iii) acetone, (iv) ethanol and acetone (1:1, v/v), and (v) acetone and ethanol (4:1, v/v) (Fig. 3.21a–e). The plate was removed when eluent front reached 25 mm from the plate top. Best TLC results were obtained in acetone and ethanol mixture (1:1, v/v) (Fig. 3.21e). Procion MX-K dye consists of three or more single near primary colours; each supplier has their own proprietary recipe. TLC showed that the colour composition of Procion MX-K consisted of near primary fibre reactive red, blue and yellow. Figure 3.21f illustrates the absorption UV-Vis spectrum, which showed that the dye predominantly consisted of blue hues. Light scattering on Fe$_3$O$_4$-paper composite and Procion MX-K dyed filter paper was determined using a reflection spectrophotometer: Both Fe$_3$O$_4$ impregnation and Procion MX-K dyeing yielded sufficient reduction in background noise (Fig. 3.21g). Both techniques can be adapted for producing lateral-flow test or multiplex assays, which can be integrated with holographic sensors. In addition to paper strips, nitrocellulose membranes were dyed using DEKA-L fabric dye. Nitrocellulose membranes were immersed in HCl

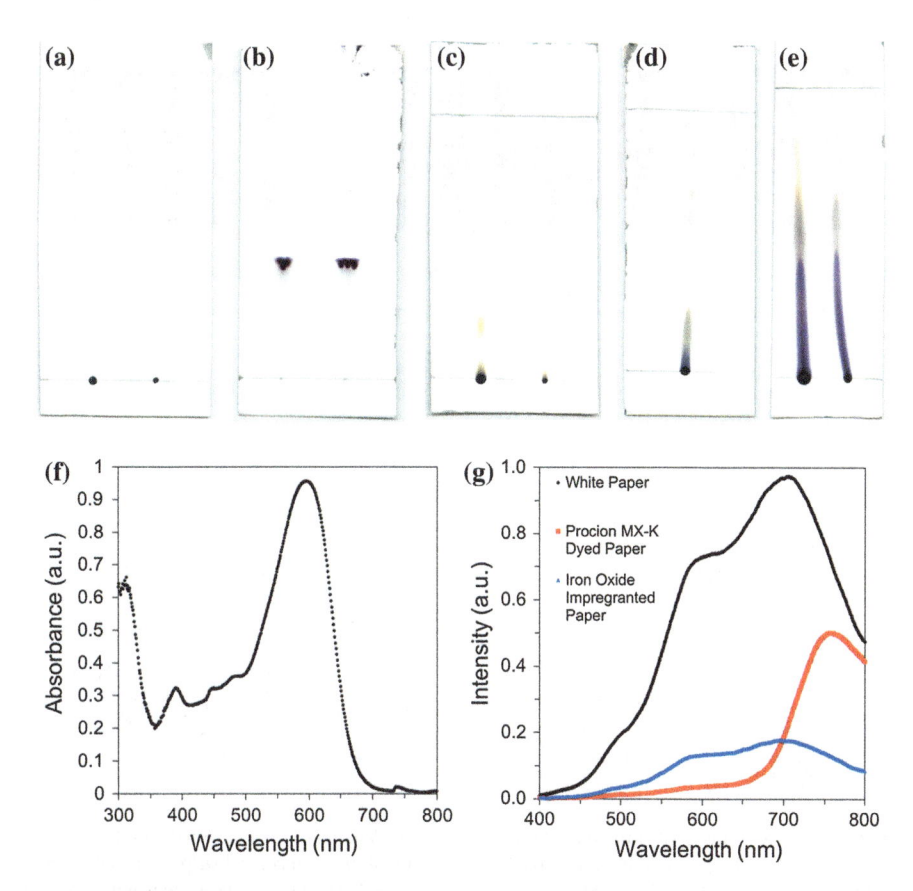

Fig. 3.21 Characterisation of Procion Black MX-K using thin-layer chromatography, and UV-Vis and reflection spectroscopy. **a** Hexane ($R_f = 0$), **b** ethanol:water (1:1, v/v) ($R_f = 0.44$), **c** acetone ($R_{f,red} = 0.22$), **d** ethanol:acetone (1:1, v/v) ($R_{f,red} = 0.22$, $R_{f,blue} = 0.04$), and **e** acetone:ethanol (4:1, v/v) ($R_{f,red} = 0.66$, $R_{f,blue} = 0.23$) at 24 °C. The *upper black lines* on the TLC plates represent the solvent front. **f** Absorption spectrum of Procion MX-K. **g** Reflection spectrum of Procion MX-K dyed filter paper, and Fe_3O_4-paper composite

(10–100 mM) while varying the temperature from 23 to 70 °C for 5 min. DEKA-L dye, sodium chloride and DI water were mixed with 2:1:40 (w/w/v). Nitrocellulose membrane samples cut to 8 mm × 2.5 cm were immersed into sealed dye baths. DEKA-L dye bath was stored at 80 °C for 2 days, and dyed nitrocellulose strips were rinsed with DI water and dried.

Mass flow rate of the paper substrates depends on the distribution of the pore size and hydrophilicity. Procion MX-K dyed/plain Whatman filter/chromatography papers were cut to 8 cm × 8 mm and attached to a levelled surface. The strips were lowered into a DI water bath with the tips immersed 3 mm and time lapse images were taken in every 10 s (Fig. 3.22a–c). The time taken to wet a specific distance was estimated (Fig. 3.22d). Capillary rise depended on the density and

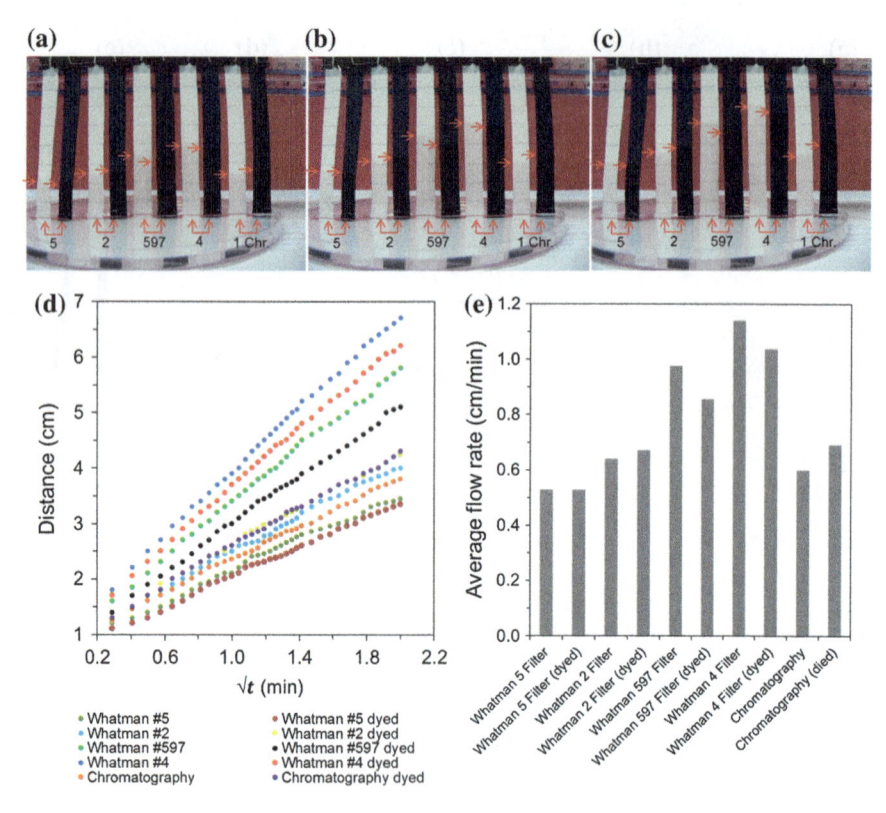

Fig. 3.22 Flow rates of Procion MX-K dyed/plain Whatman filter/chromatography papers. **a–c** Five different types of dyed test strips (8 cm × 8 mm) along with control strips were submerged into a water reservoir and the water front reach was recorded. *Red arrows* show the water front as a function of time. **d** Flow rate to wet 7 cm distance for Whatman filter/chromatography papers at 23 °C, **e** Average flow rates of the dyed papers compared to their controls

hydrophobicity of the fibres, and the viscosity of the fluid. Dyeing paper strips with Procion MX-K black slightly influenced capillary flow characteristics rendering the dyed test strips more or less hydrophobic (Fig. 3.22e).

Holographic flakes were produced by separating the polymer matrix from its substrate. Free-floating holographic sensor flakes produced visual colour changes due to variation in pH from 5.0 to 8.0 (Fig. 3.23a). Figure 3.23b illustrates holographic sensors on paper strips for monitoring pH in the metabolic range. Figure 3.23c shows the readouts for holographic pH sensor flakes in (i) free-floating, (ii) paper- and (iii) nitrocellulose-backed forms. The holographic flakes (6 mol% MAA) swelled ∼120 nm as the pH was increased from 4 to 8. This Bragg peak shift is about the half of shift obtained by the pH sensors attached to a substrate. The difference in the Bragg peak shifts might be attributed to degree of freedoms to expand. Figure 3.23d illustrates the geometry of a polymer matrix. When the polymer matrix is attached to a substrate, it can only expand normal

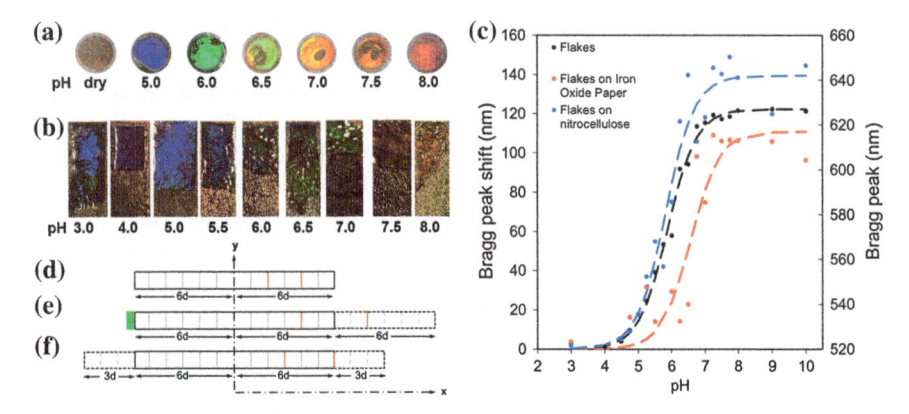

Fig. 3.23 Colorimetric readouts of holographic sensor flakes and their diffraction spectra as a function of pH values. Images of **a** free-floating flakes, **b** paper strips at different pH values in phosphate buffers (150 mM) at 24 °C. The images were taken under white light illumination. **c** Readouts for Bragg peak shifts as a function of pH change for flakes in free-floating, paper- and nitrocellulose-backed forms. **d** The reduction in the Bragg peak shift in holographic flakes. The original holographic matrix showing the positions of two individual Ag^0 NP layers (*red lines*), **e** The expansion on a plastic substrate, **f** in the flake form

Table 3.2 Apparent pK_a values of holographic pH sensors

Substrate	MAA	Glass (SH)	Glass (SR)	PMMA-backed (SR)	Flake (SR)	Paper-backed (SR)	Nitrocellulose-backed (SR)
pK_a	4.66	6.08	5.97	5.98	5.88	5.85	5.82

SH Silver halide chemistry
SR In situ size reduction of Ag^0 NPs (laser ablation)

(x direction) to its underlying substrate (shown in green) (Fig. 3.23e). On the other hand, holographic flakes can expand in both x and −x directions (Fig. 3.23f). For simplicity purposes, it was assumed that the gel does not expand in the y direction. The red lines show the position of the fringes. The expansion of the holographic matrix on a substrate is 2d, however, the expansion in the flake form is d. Table 3.2 summarises the apparent pK_a values measured from the holographic pH sensors produced by different methods.

3.5 Discussion

This chapter demonstrated the fabrication of holographic pH sensors via silver halide chemistry and in situ size reduction of Ag^0 NPs in Denisyuk reflection mode. The use of a single laser pulse (6 ns, 350 mJ) to organise Ag^0 NPs within a hydrophilic hydrogel matrix was reported. The holographic sensor was modulated to diffract narrow-band light based on the changes in Ag^0 NP spacing and index of refraction. Holographic sensors displayed reversible band gap modulation in response to variations in pH by diffracting light from the visible to the near-infrared region of the spectrum ($\lambda_{peak} \approx 495$–815 nm). The potential clinical application of the holographic sensors was demonstrated by pH sensing of artificial urine over the physiological range (4.5–9.0) with high sensitivity between pH 5.0 and 6.0. Holographic sensors allow the use of a diverse array of substrates, ranging from synthetic to natural polymers. The Bragg diffraction angle and holographic pattern within the matrix can also be controlled depending on the desired application. Additionally, other metal NPs and dyes can be used in forming well-ordered diffraction gratings with this method [24]. Holographic sensors have attractive attributes such as fabrication without requiring clean room facilities, flexible characteristics desirable for printing with potential scalability. The demonstrated fabrication strategies of holographic sensors in hydrophilic matrices may lead to applications from printable diffraction gratings to rapid colorimetric sensors.

pH sensors are employed in applications from environmental to biomedical samples. There is a range of pH sensors such as dye/molecular interaction-based sensors, fluorescent sensors, electrochemical sensors and hydrogel-based optical sensors [54]. The most commonly used sensor is based on the pH indicator colorimetric dye (Litmus paper), which is blue in a basic solution (pH > 7.0), while it is pink or red in an acidic solution (pH < 7.0) [55]. However, the colour change is not obvious (light green to dark green) for 6.5–8.0, which might cause subjective interpretation by eye. Dye-based indicators require about a minute for colour development, and they do not provide accurate results (±0.25 pH units), nevertheless they are suitable for quick measurements. The holographic sensors demonstrated in this chapter overcomes the listed issues by allowing the sensor display colour in the order from shorter to longer wavelengths in the entire visible spectrum. The present sensor can diffract colours such as magenta, cyan and yellow, unlike dye-based pH indicators. Additionally, holographic sensors allow continuous measurements. Accurate measurements can be achieved using electrochemical pH meters. These sensors consist of pH combination electrodes, where both the reference electrodes such as Ag/AgCl and a glass membrane are integrated into an electrode body [56]. The glass membrane consists of a composition that can permeate H_3O^+ ions. Electrochemical pH sensors are calibrated using three-point reference solution, and they provide rapid measurements. These sensors also have a wide working range (0.00–14.00) with a pH resolution of 0.01. However, they also have disadvantages such as (i) having a rigid design, hence mechanically fragile, (ii) being susceptible to electrical interference, and (iii) being interfered in the presence of highly alkaline solutions and fluoride ions [54]. Electrochemical sensors are also not suitable for long-term measurements due to the drift in the electrode signal. Their other drawbacks include not being suitable to measure low sample volumes. The holographic pH sensors avoid complex electronic circuitry required for electrochemical sensors, and it utilises diffraction of light, which does not require electricity to operate. In terms of accuracy, the present holographic pH sensor showed 48 nm Bragg peak shift per pH unit when read by a spectrophotometer with ± 0.5 nm resolution, which is comparable to the accuracy of electrochemical sensors. However, the sensitivity of the present sensors can be improved by reducing the concentration of the crosslinker or increasing the carboxylic acid groups in the polymer matrix [20].

In addition to colorimetric dyes and electrochemical sensing, optical sensors have been developed. Sensing platforms in this area include absorption- and luminescence-based optical systems, fibre-optic and planar waveguide-based sensors, and hydrogel-based systems that utilise diffraction and localised surface plasmon resonance [54, 57, 58]. Other sensors include optical devices coupled with pH indicators, which are based on weak organic dyes or changes on their optical properties when they are protonated or deprotonated. For example, their absorption or fluorescence properties change in the presence of acidic or basic solutions, which can be correlated with the concentration of H_3O^+ ions. Optical sensors do not require a separate reference sensor, and they can be easily miniaturised down to 1–10 µm. Additionally, they do not suffer from electromagnetic interferences, and they can be used for remote and continuous measurements. However, the drawbacks of the listed optical sensors include limited long term stability caused by photobleaching or leaching of the sensing materials while also being affected from temperature changes [54]. Notably, a change in ionic strength can alter the activity coefficients and shift of the calibration plot [59]. Hence, corrections are required to compensate for the ionic strength. Moreover, their operating pH range can be extended; indicators with multiple pK_a values or a group of receptors at different pK_a values have been used to improve their working range [60–62]. Artificial neural networks (ANN) were developed to improve their dynamic range from pH 2.0 to 12.0 [63]. Such developments included optical pH sensor arrays [64, 65]. Holographic sensors have advantages over other optical sensors since the characteristics of the diffraction grating such as the angle of Bragg angle can be controlled using arrangement of optical equipment and the nature of the laser light. Additionally, the laser writing method described in this chapter can allow patterning nanostructures such as carbon nanotubes, graphene and nanopillars [66, 67]. Laser manufacturing stands out as an efficient approach to mass produce optical sensors.

The silver-halide chemistry-based fabrication of holographic sensors requires about 10 steps, and its reduction to 2–4 steps by alternative approaches such as in situ size reduction of metal NPs can enable faster fabrication [18]. Another critical step that will accelerate the interpretation of these sensors includes response time, which should be achieved within a few seconds. The development of theoretical approaches to shorten the turnaround time will be helpful in the analysis of any hydrogel-based sensor. Holographic pH sensors developed in this chapter can be multiplexed with other holographic sensors on paper- or PDMS-based microfluidic devices or on contact lenses [68–74]. Since the readouts are colorimetric, they can be quantified by smartphones and wearable devices [75, 76]. Additionally, trials with clinical samples and comparison of the performance with commercial sensors can prove the feasibility of holographic sensing for applications in medical diagnostics, veterinary testing and environmental sensing. The development of holographic sensors will enable practical and quantitative readouts at point-of-care settings.

References

1. Kang Y, Walish JJ, Gorishnyy T, Thomas EL (2007) Broad-wavelength-range chemically tunable block-copolymer photonic gels. Nat Mater 6(12):957–960. doi:10.1038/nmat2032
2. Kim H, Ge J, Kim J, S-e Choi, Lee H, Lee H, Park W, Yin Y, Kwon S (2009) Structural colour printing using a magnetically tunable and lithographically fixable photonic crystal. Nat Photon 3(9):534–540. doi:10.1038/nphoton.2009.141
3. Kim S, Mitropoulos AN, Spitzberg JD, Tao H, Kaplan DL, Omenetto FG (2012) Silk inverse opals. Nat Photon 6(12):818–823. doi:10.1038/nphoton.2012.264
4. Li YY, Cunin F, Link JR, Gao T, Betts RE, Reiver SH, Chin V, Bhatia SN, Sailor MJ (2003) Polymer replicas of photonic porous silicon for sensing and drug delivery applications. Science 299(5615):2045–2047. doi:10.1126/science.1081298
5. Kim J, Yoon J, Hayward RC (2010) Dynamic display of biomolecular patterns through an elastic creasing instability of stimuli-responsive hydrogels. Nat Mater 9(2):159–164. doi:10.1038/nmat2606
6. Kim SO, Solak HH, Stoykovich MP, Ferrier NJ, De Pablo JJ, Nealey PF (2003) Epitaxial self-assembly of block copolymers on lithographically defined nanopatterned substrates. Nature 424(6947):411–414. doi:10.1038/nature01775
7. Ogawa S, Imada M, Yoshimoto S, Okano M, Noda S (2004) Control of light emission by 3D photonic crystals. Science 305(5681):227–229. doi:10.1126/science.1097968
8. Deubel M, von Freymann G, Wegener M, Pereira S, Busch K, Soukoulis CM (2004) Direct laser writing of three-dimensional photonic-crystal templates for telecommunications. Nat Mater 3(7):444–447. doi:10.1038/nmat1155
9. Qi M, Lidorikis E, Rakich PT, Johnson SG, Joannopoulos JD, Ippen EP, Smith HI (2004) A three-dimensional optical photonic crystal with designed point defects. Nature 429 (6991):538–542. doi:10.1038/nature02575
10. Arsenault AC, Clark TJ, von Freymann G, Cadermartiri L, Sapienza R, Bertolotti J, Vekris E, Wong S, Kitaev V, Manners I, Wang RZ, John S, Wiersma D, Ga Ozin (2006) From colour fingerprinting to the control of photoluminescence in elastic photonic crystals. Nat Mater 5 (3):179–184. doi:10.1038/nmat1588
11. Aoki K, Guimard D, Nishioka M, Nomura M, Iwamoto S, Arakawa Y (2008) Coupling of quantum-dot light emission with a three-dimensional photonic-crystal nanocavity. Nat Photon 2(11):688–692. doi:10.1038/nphoton.2008.202
12. Sibbett SS, Lopez GP (2008) Multiplex lateral flow devices and methods. US Patent 20080317633 A1 (Application)
13. Rinne SA, Garcia-Santamaria F, Braun PV (2008) Embedded cavities and waveguides in three-dimensional silicon photonic crystals. Nat Photon 2(1):52–56. doi:10.1038/nphoton.2007.252
14. Takahashi S, Suzuki K, Okano M, Imada M, Nakamori T, Ota Y, Ishizaki K, Noda S (2009) Direct creation of three-dimensional photonic crystals by a top-down approach. Nat Mater 8 (9):721–725. doi:10.1038/nmat2507
15. Ishizaki K, Noda S (2009) Manipulation of photons at the surface of three-dimensional photonic crystals. Nature 460(7253):367–370. doi:10.1038/nature08190
16. Kolle M, Lethbridge A, Kreysing M, Baumberg JJ, Aizenberg J, Vukusic P (2013) Bio-inspired band-gap tunable elastic optical multilayer fibers. Adv Mater 25(15):2239–2245. doi:10.1002/adma.201203529

17. Tsangarides CP, Yetisen AK, da Cruz Vasconcellos F, Montelongo Y, Qasim MM, Wilkinson TD, Lowe CR, Butt H (2014) Computational modelling and characterisation of nanoparticle-based tuneable photonic crystal sensors. RSC Adv 4(21):10454–10461. doi:10.1039/C3RA47984F

18. Yetisen AK, Butt H, da Cruz Vasconcellos F, Montelongo Y, Davidson CAB, Blyth J, Chan L, Carmody JB, Vignolini S, Steiner U, Baumberg JJ, Wilkinson TD, Lowe CR (2014) Light-directed writing of chemically tunable narrow-band holographic sensors. Adv Opt Mater 2(3):250–254. doi:10.1002/adom.201300375

19. Yetisen AK, Montelongo Y, Qasim MM, Butt H, Wilkinson TD, Monteiro MJ, Lowe CR, Yun SH (2014) Nanocrystal Bragg Grating Sensor for Colorimetric Detection of Metal Ions. (under review)

20. Marshall AJ, Blyth J, Davidson CA, Lowe CR (2003) pH-sensitive holographic sensors. Anal Chem 75(17):4423–4431. doi:10.1021/ac020730k

21. Jeong TH, Aumiller RW, Ro RJ, Blyth J (2002) Teaching holography workshops to beginners, practical holography XVI and holographic materials VIII, vol 4659. SPIE, San Jose, CA

22. Yoldas BE (1980) Investigations of porous oxides as an antireflective coating for glass surfaces. Appl Opt 19(9):1425–1429. doi:10.1364/AO.19.001425

23. Cariou JM, Dugas J, Martin L, Michel P (1986) Refractive-index variations with temperature of PMMA and polycarbonate. Appl Opt 25(3):334–336. doi:10.1364/AO.25.000334

24. Yetisen AK, Qasim MM, Nosheen S, Wilkinson TD, Lowe CR (2014) Pulsed laser writing of holographic nanosensors. J Mater Chem C 2(18):3569–3576. doi:10.1039/C3tc32507e

25. Blyth J (1985) Security display hologram to foil counterfeiting. In: Huff L (ed) Applications of holography. SPIE—The International Society for Optical Engineering, Los Angeles

26. Vasconcellos FD, Yetisen AK, Montelongo Y, Butt H, Grigore A, Davidson CAB, Blyth J, Monteiro MJ, Wilkinson TD, Lowe CR (2014) Printable surface holograms via laser ablation. ACS Photonics 1(6):489–495. doi:10.1021/Ph400149m

27. Asoro MA, Damiano J, Ferreira PJ (2009) Size effects on the melting temperature of silver nanoparticles: in-situ tem observations. Microsc Microanal 15:706–707. doi:10.1017/S1431927609097013

28. Martinez-Hurtado JL, Davidson CA, Blyth J, Lowe CR (2010) Holographic detection of hydrocarbon gases and other volatile organic compounds. Langmuir 26(19):15694–15699. doi:10.1021/la102693m

29. Dell'Aglio M, Gaudiuso R, ElRashedy R, De Pascale O, Palazzo G, De Giacomo A (2013) Collinear double pulse laser ablation in water for the production of silver nanoparticles. Phys Chem Chem Phys 15(48):20868–20875. doi:10.1039/c3cp54194k

30. Wagener P, Ibrahimkutty S, Menzel A, Plech A, Barcikowski S (2013) Dynamics of silver nanoparticle formation and agglomeration inside the cavitation bubble after pulsed laser ablation in liquid. Phys Chem Chem Phys 15(9):3068–3074. doi:10.1039/C2cp42592k

31. Toftmann B, Doggett B, Rgensen CBJ, Schou J, Lunney JG (2013) Femtosecond ultraviolet laser ablation of silver and comparison with nanosecond ablation. J Appl Phys 113(8). doi:10.1063/1.4792033

32. Bagratashvili VN, Rybaltovsky AO, Minaev NV, Timashev PS, Firsov VV, Yusupov VI (2010) Laser-induced atomic assembling of periodic layered nanostructures of silver nanoparticles in fluoro-polymer film matrix. Laser Phys Lett 7(5):401–404. doi:10.1002/lapl.200910159

33. Rybaltovskii AO, Gerasimova VI, Minaev NV, Sokolov VI, Timashev PS, Troitskaya EA, Firsov VV, Yusupov VI, Bagratashvili VN (2010) Laser-induced formation of structures of silver nanoparticles in fluoracrylate films impregnated with Ag(hfac)COD molecules. Nanotechnol Russ 5(7–8):435–445. doi:10.1134/S1995078010070025

34. Chung J, Han S, Lee D, Ahn S, Grigoropoulos CP, Moon J, Ko SH (2013) Nanosecond laser ablation of silver nanoparticle film. Opt Eng 52(2). doi:10.1117/1.Oe.52.2.024302

35. Po HN, Senozan NM (2001) The Henderson-Hasselbalch equation: its history and limitations. J Chem Educ 78(11):1499. doi:10.1021/ed078p1499

36. Figliola RS, Beasley DE (2011) Theory and design for mechanical measurements, 5th edn. Wiley, New York
37. Tokarev I, Minko S (2012) Tunable plasmonic nanostructures from noble metal nanoparticles and stimuli-responsive polymers. Soft Matter 8(22):5980–5987. doi:10.1039/C2sm25069a
38. Ehrick JD, Deo SK, Browning TW, Bachas LG, Madou MJ, Daunert S (2005) Genetically engineered protein in hydrogels tailors stimuli-responsive characteristics. Nat Mater 4 (4):298–302. doi:10.1038/nmat1352
39. Lendlein A, Jiang H, Junger O, Langer R (2005) Light-induced shape-memory polymers. Nature 434(7035):879–882. doi:10.1038/nature03496
40. Um SH, Lee JB, Park N, Kwon SY, Umbach CC, Luo D (2006) Enzyme-catalysed assembly of DNA hydrogel. Nat Mater 5(10):797–801. doi:10.1038/nmat1741
41. Ehrbar M, Schoenmakers R, Christen EH, Fussenegger M, Weber W (2008) Drug-sensing hydrogels for the inducible release of biopharmaceuticals. Nat Mater 7(10):800–804. doi:10.1038/nmat2250
42. Kloxin AM, Kasko AM, Salinas CN, Anseth KS (2009) Photodegradable hydrogels for dynamic tuning of physical and chemical properties. Science 324(5923):59–63. doi:10.1126/science.1169494
43. Banwell EF, Abelardo ES, Adams DJ, Ma Birchall, Corrigan A, Donald AM, Kirkland M, Serpell LC, Butler MF, Woolfson DN (2009) Rational design and application of responsive alpha-helical peptide hydrogels. Nat Mater 8(7):596–600. doi:10.1038/nmat2479
44. Stuart MAC, Huck WTS, Genzer J, Muller M, Ober C, Stamm M, Sukhorukov GB, Szleifer I, Tsukruk VV, Urban M, Winnik F, Zauscher S, Luzinov I, Minko S (2010) Emerging applications of stimuli-responsive polymer materials. Nat Mater 9(2):101–113. doi:10.1038/nmat2614
45. Yetisen AK, Montelongo Y, da Cruz Vasconcellos F, Martinez-Hurtado JL, Neupane S, Butt H, Qasim MM, Blyth J, Burling K, Carmody JB, Evans M, Wilkinson TD, Kubota LT, Monteiro MJ, Lowe CR (2014) Reusable, robust, and accurate laser-generated photonic nanosensor. Nano Lett 14(6):3587–3593. doi:10.1021/nl5012504
46. Kratz A, Ferraro M, Sluss PM, Lewandrowski KB (2004) Case records of the Massachusetts general hospital. weekly clinicopathological exercises. laboratory reference values. N Engl J Med 351(15):1548–1563. doi:10.1056/NEJMcpc049016
47. Simerville JA, Maxted WC, Pahira JJ (2005) Urinalysis: a comprehensive review. Am Fam Phys 71(6):1153–1162
48. Rodríguez Soriano J (2002) Renal tubular acidosis: the clinical entity. J Am Soc Nephrol 13 (8):2160–2170. doi:10.1097/01.ASN.0000023430.92674.E5
49. Hesse A, Heimbach D (1999) Causes of phosphate stone formation and the importance of metaphylaxis by urinary acidification: a review. World J Urol 17(5):308–315
50. Eisner BH, Goldfarb DS, Pareek G (2013) Pharmacologic treatment of kidney stone disease. Urol clin North Am 40(1):21–30. doi:10.1016/j.ucl.2012.09.013
51. Proudfoot AT, Krenzelok EP, Vale JA (2004) Position paper on urine alkalinization. J Toxicol Clin Toxicol 42(1):1–26. doi:10.1081/CLT-120028740
52. Brooks T, Keevil CW (1997) A simple artificial urine for the growth of urinary pathogens. Lett Appl Microbiol 24(3):203–206. doi:10.1046/j.1472-765X.1997.00378.x
53. Kissa E (1969) Urea in reactive dyeing. Text Res J 39(8):734–741. doi:10.1177/004051756903900805
54. Wencel D, Abel T, McDonagh C (2014) Optical chemical pH sensors. Anal Chem 86 (1):15–29. doi:10.1021/Ac4035168
55. Walpole GS (1913) The use of litmus paper as a quantitative indicator of reaction. Biochem J 7 (3):260–267
56. Bakker E, Qin Y (2006) Electrochemical sensors. Anal Chem 78(12):3965–3984. doi:10.1021/ac060637m
57. Wolfbeis OS (2008) Fiber-optic chemical sensors and biosensors. Anal Chem 80 (12):4269–4283. doi:10.1021/Ac800473b

58. Wang XD, Wolfbeis OS (2013) Fiber-optic chemical sensors and biosensors (2008–2012). Anal Chem 85(2):487–508. doi:10.1021/Ac303159b
59. Janata J (1987) Do optical sensors really measure pH? Anal Chem 59(9):1351–1356. doi:10.1021/Ac00136a019
60. Lin J, Liu D (2000) An optical pH sensor with a linear response over a broad range. Anal Chim Acta 408(1–2):49–55. doi:10.1016/S0003-2670(99)00840-5
61. Dong S, Luo M, Peng G, Cheng W (2008) Broad range pH sensor based on sol–gel entrapped indicators on fibre optic. Sens Actuators B 129(1):94–98. doi:10.1016/j.snb.2007.07.078
62. Chauhan VM, Burnett GR, Aylott JW (2011) Dual-fluorophore ratiometric pH nanosensor with tuneable pKa and extended dynamic range. Analyst 136(9):1799–1801. doi:10.1039/C1AN15042A
63. Suah FBM, Ahmad M, Taib MN (2003) Applications of artificial neural network on signal processing of optical fibre pH sensor based on bromophenol blue doped with sol–gel film. Sens Actuators B 90(1–3):182–188. doi:10.1016/S0925-4005(03)00026-1
64. Capel-Cuevas S, Cuéllar MP, de Orbe-Payá I, Pegalajar MC, Capitán-Vallvey LF (2010) Full-range optical pH sensor based on imaging techniques. Anal Chim Acta 681(1–2):71–81. doi:10.1016/j.aca.2010.09.033
65. Capel-Cuevas S, Cuéllar MP, de Orbe-Payá I, Pegalajar MC, Capitán-Vallvey LF (2011) Full-range optical pH sensor array based on neural networks. Microchem J 97(2):225–233. doi:10.1016/j.microc.2010.09.008
66. Deng S, Yetisen AK, Jiang K, Butt H (2014) Computational modelling of a graphene Fresnel lens on different substrates. RSC Adv 4(57):30050–30058. doi:10.1039/C4ra03991b
67. Kong X-T, Butt H, Yetisen AK, Kangwanwatana C, Montelongo Y, Deng S, Fd Cruz Vasconcellos, Qasim MM, Wilkinson TD, Dai Q (2014) Enhanced reflection from inverse tapered nanocone arrays. Appl Phys Lett 105(5):053108. doi:10.1063/1.4892580
68. Yetisen AK, Naydenova I, Vasconcellos FC, Blyth J, Lowe CR (2014) Holographic sensors: three-dimensional analyte-sensitive nanostructures and their applications. Chem Rev 114(20):10654–10696. doi:10.1021/cr500116a
69. Yetisen AK, Akram MS, Lowe CR (2013) Paper-based microfluidic point-of-care diagnostic devices. Lab Chip 13(12):2210–2251. doi:10.1039/c3lc50169h
70. Yetisen AK, Volpatti LR (2014) Patent protection and licensing in microfluidics. Lab Chip 14(13):2217–2225. doi:10.1039/c4lc00399c
71. Volpatti LR, Yetisen AK (2014) Commercialization of microfluidic devices. Trends Biotechnol 32(7):347–350. doi:10.1016/j.tibtech.2014.04.010
72. Akram MS, Daly R, Vasconcellos FC, Yetisen AK, Hutchings I, Hall EAH (2015) Applications of paper-based diagnostics. In: Castillo-Leon J, Svendsen WE (eds) Lab-on-a-chip devices and micro-total analysis systems. Springer, Berlin
73. Farandos NM, Yetisen AK, Monteiro MJ, Lowe CR, Yun SH (2014) Contact lens sensors in ocular diagnostics. Adv Healthc Mater. doi:10.1002/adhm.201400504
74. Yetisen AK, Jiang L, Cooper JR, Qin Y, Palanivelu R, Zohar Y (2011) A microsystem-based assay for studying pollen tube guidance in plant reproduction. J Micromech Microeng 21(5):054018. doi:10.1088/0960-1317/21/5/054018
75. Yetisen AK, Martinez-Hurtado JL, Garcia-Melendrez A, Vasconcellos FC, Lowe CR (2014) A smartphone algorithm with inter-phone repeatability for the analysis of colorimetric tests. Sens Actuators B 196:156–160. doi:10.1016/j.snb.2014.01.077
76. Yetisen AK, Martinez-Hurtado JL, da Cruz Vasconcellos F, Simsekler MC, Akram MS, Lowe CR (2014) The regulation of mobile medical applications. Lab Chip 14(5):833–840. doi:10.1039/c3lc51235e

Chapter 4
Holographic Metal Ion Sensors

The quantification of metal ions has applications in medical diagnostics, veterinary screening and environmental monitoring. This chapter describes the development of a holographic metal ion sensor through photopolymerisation. In contrast to the nanoparticles (NPs) in silver halide chemistry, porphyrin molecules were chosen for the construction of metal NP-free holographic sensors. A porphyrin derivative with acrylate groups was synthesised to crosslink 2-hydroxyethyl methacrylate monomers [1]. The porphyrin derivative also served as the light-absorbing material and cation chelating agent. A single pulse of a Nd:YAG laser ($\lambda = 532$ nm, 6 ns, 350 mJ) in Denisyuk reflection holography mode allowed formation of Bragg diffraction gratings within the porphyrin cross-linked polymer matrix. Holographic sensors had a reversible narrow-band tuneability within the visible spectrum to report on organic solvents in water as a proof of concept, and concentrations of metal cations such as Cu^{2+} and Fe^{2+} in aqueous media. The quantification of Cu^{2+} ions has a potential application in the diagnosis of Wilson's disease, a genetic disorder in which copper accumulates in the tissues [2]. Similarly, the measurement of Fe^{2+} ions may help the diagnosis of hemochromatosis, hemolytic anemia, paroxysmal nocturnal hemoglobinemia, and impaired biliary clearance [3].

The feasibility of incorporating chelating agents has been demonstrated in crystalline colloidal arrays [4–6]. In holographic sensors fabricated through the silver halide chemistry, incorporation of chelating agents into recording media was investigated to develop ion-selective hydrogel matrices [7]. To incorporate crown ethers in holographic sensors, functional groups allowing co-polymerisation of crown ethers into the polymer matrices were needed [7]. These studies involved the synthesis of methacrylate esters of homologous series of hydroxyether crown ethers and their copolymerisation with hydroxyethyl methacrylate in the presence of a crosslinker (i.e. ethylene dimethacrylate) to form a chelating hydrogel matrix. The crown ethers tests included 12-crown-4, 15-crown-5, and 18-crown-6 pendant functionalities, and they responded alkali and alkaline earth ions with varying specificity [7]. For example, holograms comprising of 18-crown-6 linearly responded to K^+ ions over the physiological range while the readouts were not affected by physiological variations in background Na^+ ion concentrations (~ 130 -150 mM). The optimised hologram containing 18-crown-6 (50 mol%) showed

A.K. Yetisen, *Holographic Sensors*, Springer Theses,
DOI 10.1007/978-3-319-13584-7_4

swelling behaviour as a function of complexation in the presence of ≤30 mM metal ions exhibiting ≤200 nm of Bragg peak shift within 30 s, showing its potential to be used in the quantification of electrolytes in biological samples. However, the selectivity of the other crown ether derivatives was shown to be limited. Another study demonstrated that a holographic sensor based on copolymers of acrylamide with ionogen comonomers was sensitive to Pb^{2+} and Co^{2+} ions (10^{-5} M), whereas the sensor's sensitivity to Mn^{2+} and Sr^{2+} ions was two orders of magnitude lower [8]. Additionally, the response to alkali metal ions (Na^+, K^+) was an order of magnitude lower than Pb^{2+} ions. Further studies in holographic sensing with chelating agents investigated incorporation of a methacrylated analogue of iminodiacetic acid (IDA), which was copolymerised with 2-hydroxyethyl methacrylate (HEMA) to form a sensor for the detection of divalent metal ions such as Ca^{2+}, Mg^{2+}, Ni^{2+}, Co^{2+} and Zn^{2+} [9]. Sensors containing >10 mol% chelating monomer and 6 mol% crosslinker shifted the Bragg peak by 46.3 nm within 30 s at an ion concentration of 0–40 mM. The relative selectivity of the holograms had a hierarchy of Ni^{2+}> Zn^{2+} > Co^{2+} > Ca^{2+} > Mg^{2+} ions. The real-time reversible response of the sensor was demonstrated in monitoring Ca^+ ion efflux during the early stages of germination of *Bacillus megaterium* spores [9].

4.1 Fabrication of Holographic Metal Ion Sensors via Photopolymerisation

Porphyrins have a versatile nature in coordination with the analytes and synthetic modularity [10]. They have been used in organic solar cells [11], non-linear optics [12], catalysis [13] and odour visualisation [14]. In the present study, a porphyrin derivative was used as a dye pigment in the fabrication of the holographic sensor. To achieve homogenous solubility, these molecules have been further modified with acrylate groups to serve as a crosslinker as well as laser-light interactive pigments. Tetra carboxyphenyl porphyrin (TACPP) **1** was synthesised as reported previously [15], and it was further condensed with 3-(4-hydroxy-phenoxy)propyl acrylate to obtain the desired product (Fig. 4.1). *p*-Carboxybenzaldehyde (4.00 g, 26.5 mmol) was mixed in propanoic acid (∼200 mL) in a round-bottom flask fitted with a condenser. The reaction mixture was heated under reflux for 1 h followed by the successive dropwise addition of pyrrole (1.9 mL, 26.5 mmol) via a syringe. The resultant dark mixture was refluxed with continued stirring for ~3 h under a constant flow of air. The product was separated from the reaction mixture by hot filtration and washed with warm dichloromethane (DCM) followed by a small amount of cold methanol. The filtrate was collected, dried under vacuum and purified by recrystallisation from methanol/DCM (50:50, v/v), desired product **1** was obtained as a purple solid (Fig. 4.1). Yield: 1.5 g, 28 %. ^{1}H NMR (500 MHz, DMSO): δ -2.94 (s, 2H, NH), 7.30−7.32(m, 8H, ArH). 8.37−8.41 (m, 8H, ArH). 8.65 (s, 8H, pyrrolic-β-CH), and 13.32 (s, 4H, COOH). FT-IR (cm^{-1}): 3435, 3061, 1685, 1600, 1285, 782. TACPP **1**

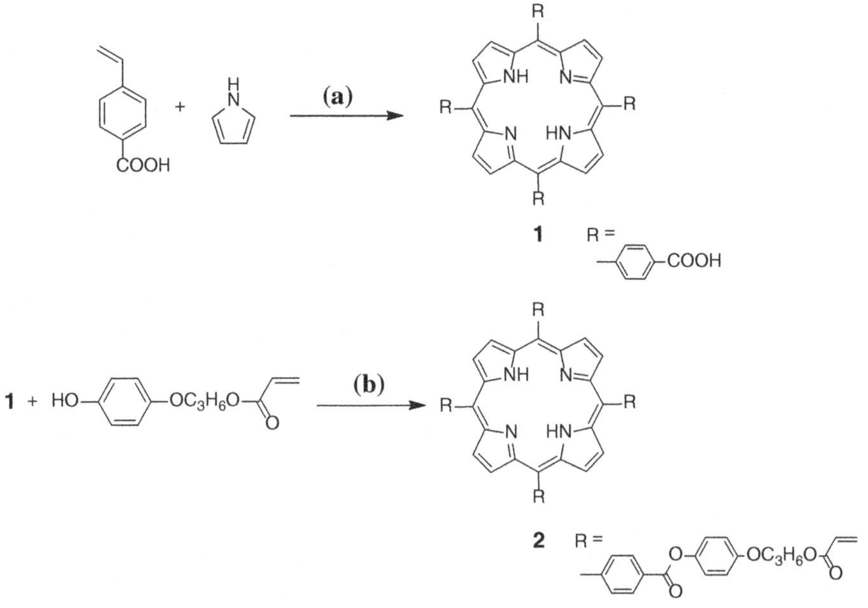

Fig. 4.1 Synthetic scheme of TACPP 2 **a** propanoic acid, reflux 3 h and **b** dichloromethane, DCC, DMAP, stir room temp., 48 h. Reproduced from Ref. [1] with permission from The Royal Society of Chemistry

(1.00 g, 1.2 mmol) was dissolved in cold DCM ($\sim$100 mL) followed by N,N'-dicyclohexylcarbodiimide (DCC) (1.06 g, 5.1 mmol) and 4-(dimethylamino)pyridine (DMAP) (0.62 g, 5.1 mmol) addition. The resultant mixture was stirred below room temp (<24 °C) for 1 h followed by the addition of 3-(4-hydroxyphenoxy) propyl acrylate (1.13 g, 5.1 mmol). The mixture was left in dark under constant stirring and nitrogen atmosphere for 48 h. The reaction mixture was then filtered, evaporated under reduced pressure and purified by flash column chromatography eluted with DCM on silica gel. The major fraction was a deep red colour and was evaporated to dryness to yield a shiny purple solid of 5,10,15,20-tetrakis[4″-(3‴-(acryloyloxy) propoxy)phenyl-4′-car-boxyphenyl] porphyrin (TACPP) **2**.

A monomer solution consisting of HEMA (99.5 mol%) and TACPP **2** (0.5 mol %) in tetrahydrofuran (THF) were mixed (1:1, v/v) with 2-dimethoxy-2-pheny-lacetophenone (DMPA, 2 %, w/v) in THF. Poly(methyl methacrylate) (PMMA) substrate surface was treated with O_2 plasma under a vacuum (1 torr, 30 s) to render the PMMA surface hydrophilic. The monomer mixture ($\sim$100 μl) was deposited on an Al coated polyester sheet. The PMMA substrates, O_2-plasma-treated side facing down, were positioned on the solution. Silica beads ($\sim\varnothing$10 μm) were used to control the thickness of the monomer mixture. The samples were subsequently exposed to UV light (λ = 350 nm) for 20 min to initiate free-radical polymerisation, in which TACPP **2** served as the cross linker. The final polymer matrix had a thickness of 10 ± 2 μm. Higher concentrations (1−2 mol%) of TACPP **2** may be preferred for designing more rigid polymers. Fabrication of diffraction grating consisted of passing a laser beam through a pHEMA-co-TACPP matrix, which was aligned with an object (i.e. planar mirror) behind. In "Denisyuk" reflection mode, incidence and reflected beams formed constructive wave interferences [16, 17]. The pHEMA-co-TACPP matrix was selectively patterned at the antinodes of the standing wave, where the energy of the laser light was maximum, to form a photonic structure, which acted as a narrow-band Bragg reflector. The laser energy absorbed by the pHEMA-co-TACPP matrix depends on the laser wavelength, power, pulse duration, dye's thermal (absorption), physical and chemical characteristics. The photonic structure consisted of a periodic grating organised roughly half of the wavelength of the laser light that was used to produce the grating. When the system was exposed to the laser light, the pHEMA matrix was further cross-linked with the help of the TACPP molecules at the antinodes of the standing wave. UV-Vis absorption spectrum shows TACPP **2** molecules in THF with peaks at 422 nm (Soret band) and 453 nm (Fig. 4.2a). UV-Vis absorption spectra showed that TACPP **2** molecules covalently bound to pHEMA matrix before and after laser-induced photochemical patterning (Fig. 4.2a). The lattice spacing and the refractive-index contrast of the sensor was modulated by external stimuli, which systematically shifted the Bragg peak from shorter to longer wavelengths when the hydrogel expanded in the direction normal to underlying substrate (Fig. 4.2b).

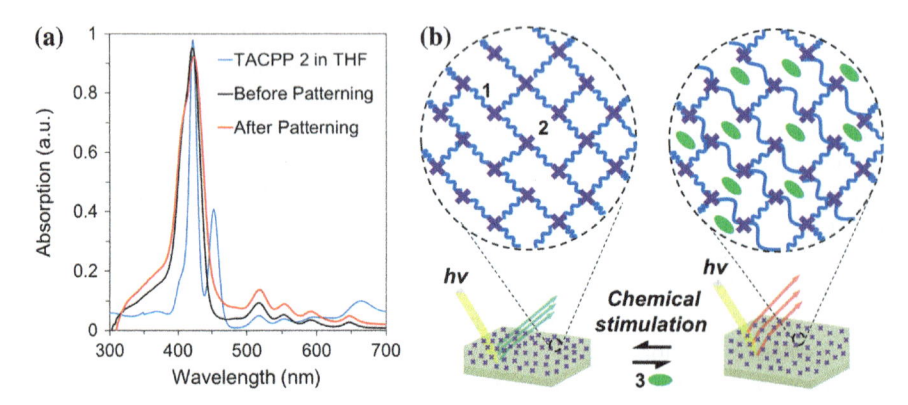

Fig. 4.2 Characterisation and principle of operation of a porphyrin-functionalised holographic sensor. **a** TACPP 2 in UV-Vis spectroscopy showing peaks at 422 nm (Soret band) and 453 nm. TACPP 2 molecules covalently bound to pHEMA matrix before and after laser-light exposure. **b** The sensor comprised pHEMA (**1**) crosslinked with TACPP (**2**) was immediately ready to sense the target (**3**). Swelling of the sensor in response to external stimuli shifted the Bragg peak from shorter to longer wavelengths as the pHEMA-co-TACPP matrix expanded normal to its underlying substrate. Reproduced from Ref. [1] with permission from The Royal Society of Chemistry

4.2 Optical Readouts

4.2.1 Organic Solvents in Water

To demonstrate the tuneability of the porphyrin-based sensor, its response to ethanol, methanol, propan-2-ol and dimethyl sulphoxide (DMSO) in DI water (0.0−10.0 %, v/v) was investigated. An increase in the solvent concentrations systematically swelled the hydrogel matrix. Figure 4.3a shows a typical Bragg peak shift from shorter ($\sim$520 nm) to longer ($\sim$620 nm) wavelengths as a function of solvent concentration (e.g. ethanol). As the lattice spacing increased, the diffraction efficiency of the periodic structure decreased. This may be attributed to the decrease in the effective refractive index contrast in the expanding hydrogel matrix. The expanding photonic structure displayed a colorimetric response in the visible spectrum throughout the measurement (e.g. ethanol, Fig. 4.3b). The sensor response to solvents was fully reversible and rapid with 90 % of the equilibrium response measured within 1 min. Figure 4.3c illustrates the Bragg peak shifts as a function of ethanol, methanol, propan-2-ol and DMSO concentrations over three trials, which exhibited ±3 % variation. The Bragg peak shifts for 10.0 % (v/v) ethanol, methanol, propan-2-ol and DMSO were 100, 48, 162 and 56 nm, respectively. Consecutive swelling/shrinking processes were reproducible over 30 successive changes without hysteresis. Response time and recovery times for 10.0 % (v/v) ethanol, methanol, propan-2-ol and DMSO were $\sim$50, $\sim$60 s; $\sim$30, $\sim$40 s; $\sim$50, $\sim$50 s; and $\sim$10, $\sim$20 s, respectively (Fig. 4.3d). The water-alcohol solutions penetrated into the hydrogel matrix and caused swelling, which increased with the alkyl chain length of the alcohol [18]. In this respect, for example, water-ethanol mixture was more effective than water-methanol mixture of the same concentration. This may be attributed to an increase in hydrophilicity due to the increase in alkyl chain length. The sensor response to pure organic solvents was also investigated. Bragg peak shifts were recorded at −38, −24 and 37 nm from a starting peak wavelength of $\sim$520 nm for pure DCM, chloroform and THF, respectively. DCM and chloroform caused a contraction. This may be attributed to the capacity of DCM and chloroform to expel the water molecules (RH 60 %) from the pHEMA matrix.

4.2.2 Quantification of Cu^{2+} and Fe^{2+} Ions in Aqueous Solutions

The pK_a values of mesoporphyrins are <6.5 [19]. Upon changing the pH of the pHEMA-co-TACPP matrix from 7.0 to 3.0, the absorption peak shifted from 420 nm (pH 7.0) to 425 nm (pH 3.0) (Fig. 4.4a). The apparent pK_a values of the pHEMA-co-TACPP matrix were $\sim$3.14 and $\sim$6.59 determined using the Henderson-Hasselbalch equation (Fig. 4.4b). A typical porphyrin molecule has four pyrrole rings linked via methine bridges, resulting in a stable structure and aromatic

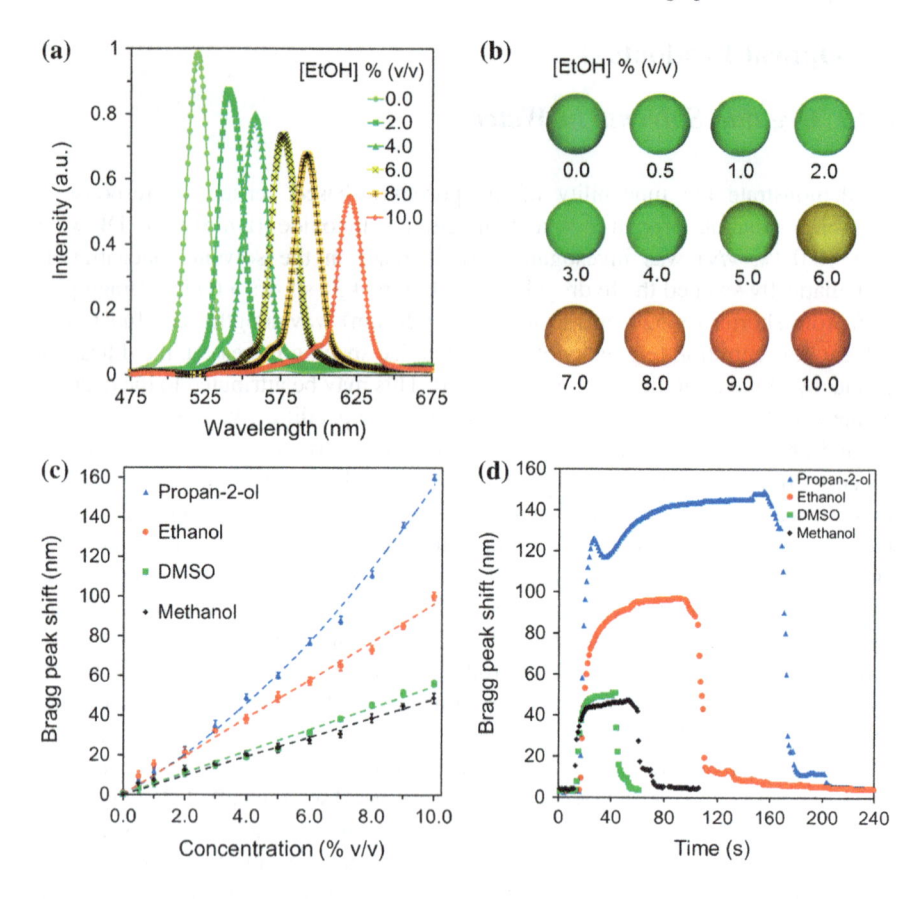

Fig. 4.3 Modulation of the porphyrin-functionalised holographic sensor due to external stimuli. **a** Bragg peak shift due to an increase in the solvent (e.g. ethanol) concentration from shorter to longer wavelengths ($\sim$520–620 nm), **b** Colorimetric response to a variation in ethanol concentration (10.0 % v/v), **c** Sensor response as a function of ethanol, methanol, propan-2-ol and DMSO concentrations over three trials, **d** Response time and reversibility. Organic solvents (10.0 %, v/v) were introduced to the holographic sensor, and replaced by DI water. Reproduced from Ref. [1] with permission from The Royal Society of Chemistry

character (Fig. 4.4c) [20]. The nucleus of the porphyrin molecule is a tetradentate ligand, which allows chelating a coordinated metal with a diameter of $\sim$3.7 Å [21]. When a cation binds to the cavity, two protons are removed from the pyrrole nitrogen atoms and two nitrogens share their lone pair of electrons, resulting in complexation. Hence, to increase the chelation, the cavity of the porphyrin was deprotonated by increasing the pH above 7.0 (Na_2CO_3–$NaHCO_3$ buffer at pH 9.2, 100 mM, 24 °C), followed by a rinse with DI water (Fig. 4.4d).

Porphyrins are macrocyclic chelating agents for multivalent metal cations. Most divalent metal ions form 1:1 complexes with porphyrins. Chelation with divalent metal ions results in a tetracoordinate chelate with no residual charge, where metal

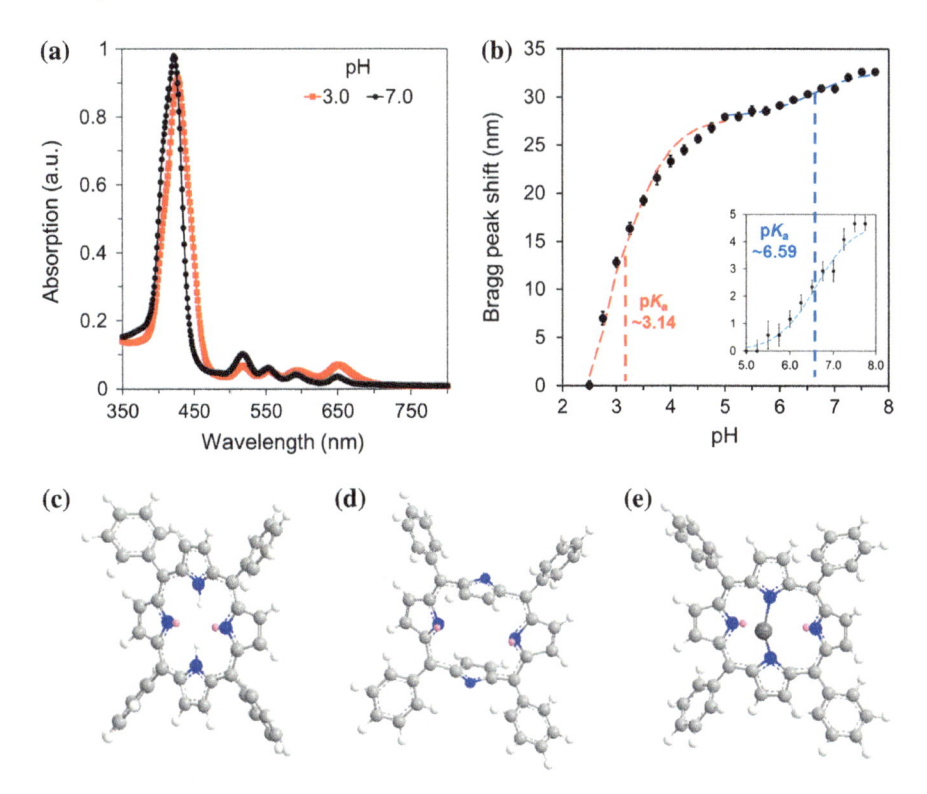

Fig. 4.4 Determination of the apparent pK_a value and activation of TACPP cavity. **a** Ultraviolet-visible spectra of pHEMA-co-TACPP matrix.As the pH was changed from 7.0 to 3.0, the absorption peak shifted from 420 to 426 nm. **b** The apparent pK_a values for the porphyrin-crosslinked pHEMA matrix were calculated using the Henderson-Hasselbalch equation. TACPP states in **c** neutral, **d** deprotonated, and **e** chelated with a divalent cation. Reproduced from Ref. [1] with permission from The Royal Society of Chemistry

atoms are located below and above the macrocycle plane of the porphyrin (Fig. 4.4e) [20]. After the cavity of TACPP was deprotonated, the pHEMA-co-TACPP matrix was saturated with mono/divalent ions (1.0 M) in buffer solutions. In response to metal cations, the Bragg peak of the pHEMA-co-TACPP matrix blue shifted (Fig. 4.5a, b). The shrinkage in the matrix was due to (i) electrostatic interactions, which decreased the Donnan osmotic pressure, and (ii) TACPP and metal cation chelation that changed the conformation of the molecule. By rinsing the system with DI water, the electrostatic interaction was eliminated, which allowed the measurement of the relative blue Bragg shift due to the chelation (Fig. 4.5c). The blue Bragg shifts for Cu^{2+} and Fe^{2+} ions were 5.24 and 4.66 nm, which were more pronounced than other mono/divalent cations. When the system was rinsed with the carbonate buffer and pure DI water, the Bragg peak shifted back to its original position (~ 523 nm). In the readouts, the absorption of Cu^{2+} and Fe^{2+} ions in the polymer matrix was negligible as compared to the Bragg peak shift.

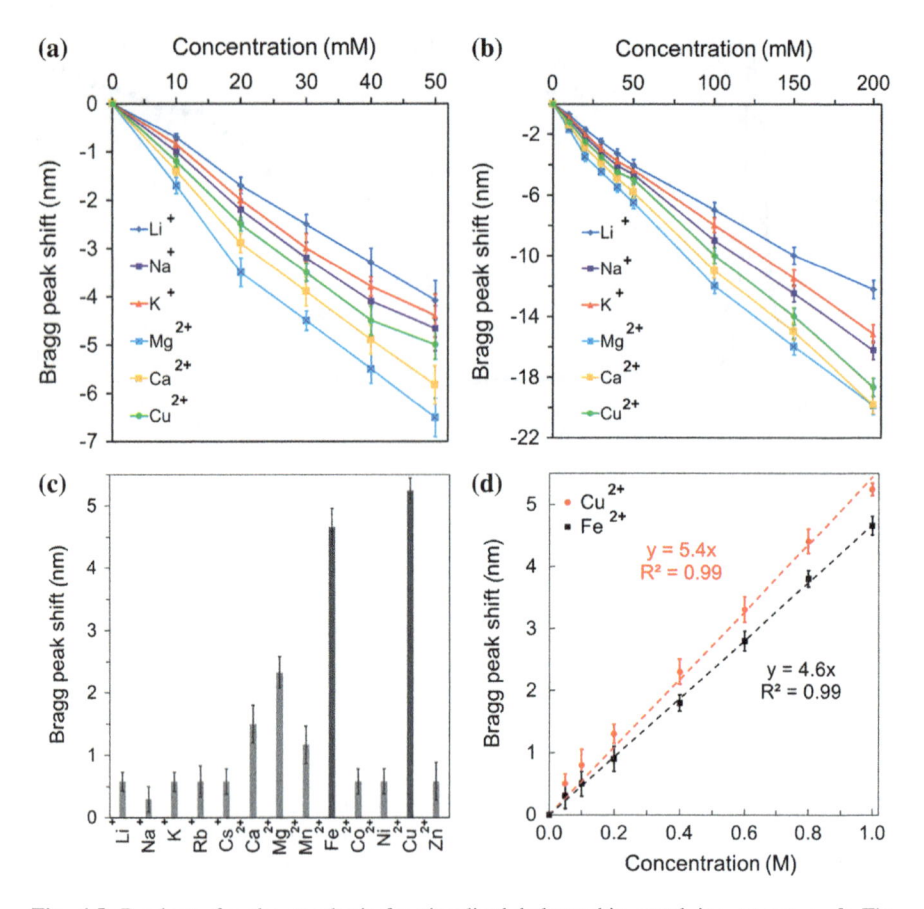

Fig. 4.5 Readouts for the porphyrin-functionalised holographic metal ion sensor. **a, b** The shrinkage of the pHEMA-co-TACPP matrix due to Donnan osmotic pressure (dominant) and the chelation effect at 24 °C in citric acid-Na_2HPO_4 buffers. **c** Blue Bragg shifts due to metal cation chelation. The sensor displays a higher blue Bragg shifts for Cu^{2+} and Fe^{2+} cations than monovalent and other divalent cations at 1.0 M. **d** Blue Bragg shifts as a function of Cu^{2+} and Fe^{2+} cation concentrations. Reproduced from Ref. [1] with permission from The Royal Society of Chemistry

The sensor's response to Cu^{2+} and Fe^{2+} ions with 200 mM increments was measured over three trials (Fig. 4.5d). The stability constant of the cavity of TACPP with cations depends on the chelate effect (entropy) [22], macrocyclic effect [23], geometrical factors [24], classification of donor atoms [25] and ionic radius [26]. For example, Cu^{2+} ions have a greater capacity to bind to the cavity of TACPP than other cations [26]. In terms of response time, for example, a variation in Cu^{2+} ion concentration from 0 to 200 mM, followed by the rinsing step, required ~ 30 s to equilibrate ($\pm$sub nm). The sensor response was reversible without hysteresis.

Table 4.1 Recent approaches to metal ion sensing

Sensing mechanism	References
Fluorescent chemosensors	[41–52]
DNA-based detection	[53–60]
Paper-based electrochemical sensors	[61]
Silica nanotube-based sensors	[62]
Optical cage sensors	[63, 64]
Optical nanosensors	[65]
Crystalline colloidal arrays	[5, 66]
Plasmonic resonance energy transfer-based nanospectroscopy	[67]
Quantum-dot-labeled DNAzymes	[68]
Polydiacetylene-liposome microarrays	[69]
Catalytic nanomotors	[70]
Click chemistry-based detection	[71]
Graphene-based sensors	[72–74]

4.3 Conclusions

Metal cation sensors have a wide array of applications from in vitro diagnostics, assessing the quality of water to intracellular sensing [27–32]. Advanced techniques for quantifying concentrations of metal cations are based on high resolution inductively coupled plasma mass spectrometry and atomic absorption or emission spectroscopy [33–35]. However, these techniques are high cost, require significant instrumentation and are non-portable. In contrast, the holographic sensors demonstrated in this chapter offers miniaturisation, however, it does not overcome the problem of equipment requirement since the readouts are taken by spectro-photometers. Other methods involve quantifying the concentration of metal cations using fluorescence [36, 37]. Such techniques do not require high-temperature atomisation sources, but require sample preparation involving organic solvents to allow binding of complexation agents to cations. Holographic metal ion sensors overcome this limitation since they do not require complex sample preparation steps. Practical solutions involve ion-selective electrodes (ISEs) that can provide high selectivity and sensitivity for the detection metal ions such as Na^+, K^+, Pb^{2+} and Cd^{2+} [38, 39], while being amenable to miniaturisation in battery-operated devices. Although electrochemical sensors require reference solutions and custom readout equipment to calibrate the electrodes, holographic sensors do not require calibration. The need for improving the performance has motivated the investigation of numerous other approaches (Table 4.1). These studies have achieved quantification of metal cations at trace concentrations. However, the attributes of being colorimetric, label/equipment-free, low-cost, lightweight, reusable or disposable in a single metal cation sensor; while also being amenable to mass manufacturing remains a significant challenge. Improvement of such technologies can enable rapid point-of-care diagnostics, particularly in the developing world, where rapid tests are not affordable. For example, Abbott Laboratories markets the i-STAT system, a handheld device that integrates microfluidics and electrochemical detection to analyse blood chemistry [40]. i-STAT quantifies analytes such as electrolytes, metabolites, and gases, while also having the capability to perform immunoassays. However, its high price and dependence on cartridges limits its potential for low-cost rapid diagnostics.

In this chapter, a porhyrin derivative was synthesised to produce a NP-free holographic metal ion sensor. The porphyrin molecule served purposes such as crosslinking of the HEMA monomers, light absorption during grating fabrication, as well as chelating agent for cation sensing. The holographic sensor was fabricated with a fraction of time and complexity when as compared to other grating fabrication techniques [75–78]. The sensor was used to quantify the concentrations of organic solvents in water, and used porphyrin as a chelating agent to quantify Cu^{2+} and Fe^{2+} ions in solutions. The sensor diffracted narrow-band light in the visible region enabling visual readouts for solvent measurements in water. However, the sensor was insensitive to low concentration of metal cations since the amount of porphyrin derivative in the polymer matrix served as the chelating agent as well as the cross-linker, which limited the swelling of the polymer matrix. The design of pendant porphyrin derivatives may increase the sensitivity and allow the pHEMA matrix swell to enable colorimetric readouts in the entire visible spectrum. In terms of fabrication flexibility, the diffraction angle and the holographic pattern (e.g. photo-masks) can be controlled depending on the desired application. Additionally, the laser writing technique described in this chapter can be used to pattern nanostructures on various surfaces [79, 80]. Holography also allows fabrication and printing of sensors in three-dimensional networks of hydrophobic materials (e.g. poly(dimethylsiloxane) (PDMS)) and polyacrylamide or hybrid polymers [81–84]. Holographic sensors can be functionalised to be sensitive to many analytes such as pH, metal ions, glucose, lactate and fructose [85–89]. It is envisioned that holographic sensors will be incorporated in multiplex microfluidic devices, contact lenses and wearable devices [40, 90–93]. Since holographic sensors are colorimetric, they might be quantified using smartphones and smartwatches [94, 95]. Holographic sensing is label-free, which not only serves as an analyte receptor, but also an optical transducer for colorimetric readouts. The porphyrin-based holographic sensor may find applications from environmental monitoring to biochemical detection.

References

1. Yetisen AK, Qasim MM, Nosheen S, Wilkinson TD, Lowe CR (2014) Pulsed laser writing of holographic nanosensors. J Mater Chem C 2(18):3569–3576. doi:10.1039/C3tc32507e
2. Ala A, Walker AP, Ashkan K, Dooley JS, Schilsky ML (2007) Wilson's disease. Lancet 369 (9559):397–408. doi:10.1016/S0140-6736(07)60196-2

3. Test ID: FEU, Iron, 24 Hour, Urine. Mayo Clinic. http://www.mayomedicallaboratories.com. Accessed 27 Oct 2014

4. Holtz JH, Asher SA (1997) Polymerized colloidal crystal hydrogel films as intelligent chemical sensing materials. Nature 389(6653):829–832. doi:10.1038/39834

5. Asher SA, Sharma AC, Goponenko AV, Ward MM (2003) Photonic crystal aqueous metal cation sensing materials. Anal Chem 75(7):1676–1683. doi:10.1021/ac026328n

6. Baca JT, Finegold DN, Asher SA (2008) Progress in developing polymerized crystalline colloidal array sensors for point-of-care detection of myocardial ischemia. Analyst 133(3):385–390. doi:10.1039/B712482a

7. Mayes AG, Blyth J, Millington RB, Lowe CR (2002) Metal ion-sensitive holographic sensors. Anal Chem 74(15):3649–3657. doi:10.1021/ac020131d

8. Kraiskii AV, Postnikov VA, Sultanov TT, Khamidulin AV (2010) Holographic sensors for diagnostics of solution components. IEEE J Quantum Electron 40(2):178–182. doi:10.1070/Qe2010v040n02abeh014169

9. Gonzalez BM, Christie G, Davidson CAB, Blyth J, Lowe CR (2005) Divalent metal ion-sensitive holographic sensors. Anal Chim Acta 528(2):219–228. doi:10.1016/j.aca.2004.03.029

10. Tonezzer M, Maggioni G, Dalcanale E (2012) Production of novel microporous porphyrin materials with superior sensing capabilities. J Mater Chem 22(12):5647–5655. doi:10.1039/C2jm15008e

11. Wöhrle D, Meissner D (1991) Organic solar cells. Adv Mater 3(3):129–138. doi:10.1002/adma.19910030303

12. Rao DVGLN, Aranda FJ, Roach JF, Remy DE (1991) Third-order, nonlinear optical interactions of some benzporphyrins. Appl Phys Lett 58(12):1241–1243. doi:10.1063/1.104323

13. Collman JP, Halbert TR, Suslick KS (1980) Oxygen binding to heme proteins and their synthetic analogs. Met Ions Biol 2:1–72

14. Rakow NA, Suslick KS (2000) A colorimetric sensor array for odour visualization. Nature 406 (6797):710–713. doi:10.1038/35021028

15. Twyman LJ, Ellis A, Gittins PJ (2011) Synthesis of multiporphyrin containing hyperbranched polymers. Macromolecules 44(16):6365–6369. doi:10.1021/Ma200863u

16. Saxby G (2004) Practical holography, 3rd edn. Institute of Physics Publishing, London

17. Benton SA, Bove VM (2007) In-line "Denisyuk" reflection holography. In: Holographic imaging. Wiley, USA. doi:10.1002/9780470224137.ch16

18. Mayes AG, Blyth J, Kyrolainen-Reay M, Millington RB, Lowe CR (1999) A holographic alcohol sensor. Anal Chem 71(16):3390–3396. doi:10.1021/Ac990045m

19. Neuberger A, Scott JJ (1952) The basicities of the nitrogen atoms in the porphyrin nucleus; their dependence on some substituents of the tetrapyrrolic ring. Philos Trans R Soc A 213(1114):307–326. doi:10.1098/rspa.1952.0128

20. Biesaga M, Pyrzyńska K, Trojanowicz M (2000) Porphyrins in analytical chemistry. A review. Talanta 51(2):209–224. doi:10.1016/S0039-9140(99)00291-X

21. Bhushan B (2009) Biomimetics: lessons from nature—an overview. Philos Trans R Soc A 367(1893):1445–1486. doi:10.1098/rsta.2009.0011

22. Schwarzenbach G (1952) Der Chelateffekt. Helv Chim Acta 35(7):2344–2359. doi:10.1002/hlca.19520350721

23. Cabbiness DK, Margerum DW (1969) Macrocyclic effect on the stability of copper(II) tetramine complexes. J Am Chem Soc 91(23):6540–6541. doi:10.1021/ja01051a091

24. Lundeen M, Hugus ZZ (1992) A calorimetric study of some metal ion complexing equilibria. Thermochim Acta 196(1):93–103. doi:10.1016/0040-6031(92)85009-K

25. Ahrland S, Chatt J, Davies NR (1958) The relative affinities of ligand atoms for acceptor molecules and ions. Q Rev Chem Soc 12(3):265–276. doi:10.1039/QR9581200265

26. Irving H, Williams RJP (1953) 637. The stability of transition-metal complexes. J Chem Soc:3192–3210. doi:10.1039/JR9530003192

27. Waldron KJ, Rutherford JC, Ford D, Robinson NJ (2009) Metalloproteins and metal sensing. Nature 460(7257):823–830. doi:10.1038/nature08300

28. Aragay G, Pons J, Merkoci A (2011) Recent trends in macro-, micro-, and nanomaterial-based tools and strategies for heavy-metal detection. Chem Rev 111(5):3433–3458. doi:10.1021/cr100383r

29. Jung JH, Lee JH, Shinkai S (2011) Functionalized magnetic nanoparticles as chemosensors and adsorbents for toxic metal ions in environmental and biological fields. Chem Soc Rev 40 (9):4464–4474. doi:10.1039/C1cs15051k

30. Kim HN, Ren WX, Kim JS, Yoon J (2012) Fluorescent and colorimetric sensors for detection of lead, cadmium, and mercury ions. Chem Soc Rev 41(8):3210–3244. doi:10.1039/C1cs15245a

31. Albelda MT, Frias JC, Garcia-Espana E, Schneider HJ (2012) Supramolecular complexation for environmental control. Chem Soc Rev 41(10):3859–3877. doi:10.1039/c2cs35008d

32. Dudev T, Lim C (2014) Competition among metal ions for protein binding sites: determinants of metal ion selectivity in proteins. Chem Rev 114(1):538–556. doi:10.1021/Cr4004665

33. Bings NH, Bogaerts A, Broekaert JA (2010) Atomic spectroscopy: a review. Anal Chem 82(12):4653–4681. doi:10.1021/ac1010469

34. Profrock D, Prange A (2012) Inductively coupled plasma-mass spectrometry (ICP-MS) for quantitative analysis in environmental and life sciences: a review of challenges, solutions, and trends. Appl Spectrosc 66(8):843–868. doi:10.1366/12-06681

35. Liu R, Wu P, Yang L, Hou X, Lv Y (2013) Inductively coupled plasma mass spectrometry-based immunoassay: a review. Mass Spectrom Rev 9999:1–21. doi:10.1002/mas.21391

36. Lodeiro C, Capelo JL, Mejuto JC, Oliveira E, Santos HM, Pedras B, Nunez C (2010) Light and colour as analytical detection tools: a journey into the periodic table using polyamines to bio-inspired systems as chemosensors. Chem Soc Rev 39(8):2948–2976. doi:10.1039/B819787n

37. Zhao Q, Li F, Huang C (2010) Phosphorescent chemosensors based on heavy-metal complexes. Chem Soc Rev 39(8):3007–3030. doi:10.1039/b915340c

38. Bobacka J, Ivaska A, Lewenstam A (2008) Potentiometric ion sensors. Chem Rev 108 (2):329–351. doi:10.1021/cr068100w

39. Dimeski G, Badrick T, St John A (2010) Ion selective electrodes (ISEs) and interferences-A review. Clinica Chimica Acta 411(5−6):309−317. doi:10.1016/j.cca.2009.12.005

40. Volpatti LR, Yetisen AK (2014) Commercialization of microfluidic devices. Trends Biotechnol 32(7):347–350. doi:10.1016/j.tibtech.2014.04.010

41. Wegner SV, Okesli A, Chen P, He C (2007) Design of an emission ratiometric biosensor from MerR family proteins: a sensitive and selective sensor for Hg^{2+}. J Am Chem Soc 129 (12):3474–3475. doi:10.1021/ja068342d

42. Huang CC, Yang Z, Lee KH, Chang HT (2007) Synthesis of highly fluorescent gold nanoparticles for sensing mercury(II). Angew Chem 46(36):6824–6828. doi:10.1002/anie.200700803

43. Nolan EM, Lippard SJ (2007) Turn-on and ratiometric mercury sensing in water with a red-emitting probe. J Am Chem Soc 129(18):5910–5918. doi:10.1021/ja068879r

44. Zhang XA, Lovejoy KS, Jasanoff A, Lippard SJ (2007) Water-soluble porphyrins as a dual-function molecular imaging platform for MRI and fluorescence zinc sensing. Proc Natl Acad Sci USA 104(26):10780–10785. doi:10.1073/pnas.0702393104

45. Cheng T, Xu Y, Zhang S, Zhu W, Qian X, Duan L (2008) A highly sensitive and selective OFF-ON fluorescent sensor for cadmium in aqueous solution and living cell. J Am Chem Soc 130(48):16160–16161. doi:10.1021/ja806928n

46. Taki M, Desaki M, Ojida A, Iyoshi S, Hirayama T, Hamachi I, Yamamoto Y (2008) Fluorescence imaging of intracellular cadmium using a dual-excitation ratiometric chemosensor. J Am Chem Soc 130(38):12564–12565. doi:10.1021/Ja803429z

47. Zhang XA, Hayes D, Smith SJ, Friedle S, Lippard SJ (2008) New strategy for quantifying biological zinc by a modified zinpyr fluorescence sensor. J Am Chem Soc 130(47):15788–15789. doi:10.1021/ja807156b

48. Ye BC, Yin BC (2008) Highly sensitive detection of mercury(II) ions by fluorescence polarization enhanced by gold nanoparticles. Angew Chem 47(44):8386–8389. doi:10.1002/anie.200803069

49. Tomat E, Nolan EM, Jaworski J, Lippard SJ (2008) Organelle-specific zinc detection using zinpyr-labeled fusion proteins in live cells. J Am Chem Soc 130(47):15776–15777. doi:10.1021/Ja806634e

50. Chatterjee A, Santra M, Won N, Kim S, Kim JK, Kim SB, Ahn KH (2009) Selective fluorogenic and chromogenic probe for detection of silver ions and silver nanoparticles in aqueous media. J Am Chem Soc 131(6):2040–2041. doi:10.1021/ja807230c

51. Marbella L, Serli-Mitasev B, Basu P (2009) Development of a fluorescent Pb^{2+} sensor. Angew Chem Int Ed 48(22):3996–3998. doi:10.1002/anie.200806297

52. Huang L, Hou FP, Xi P, Bai D, Xu M, Li Z, Xie G, Shi Y, Liu H, Zeng Z (2011) A rhodamine-based "turn-on" fluorescent chemodosimeter for Cu^{2+} and its application in living cell imaging. J Inorg Biochem 105(6):800–805. doi:10.1016/j.jinorgbio.2011.02.012

53. Lee JS, Han MS, Mirkin CA (2007) Colorimetric detection of mercuric ion (Hg^{2+}) in aqueous media using DNA-functionalized gold nanoparticles. Angew Chem 46(22):4093–4096. doi:10.1002/anie.200700269

54. Li T, Dong SJ, Wang E (2009) Label-free colorimetric detection of aqueous mercury ion (Hg^{2+}) using Hg^{2+}-Modulated G-Quadruplex-Based DNAzymes. Anal Chem 81(6):2144–2149. doi:10.1021/Ac900188y

55. Wang H, Kim Y, Liu H, Zhu Z, Bamrungsap S, Tan W (2009) Engineering a unimolecular DNA-catalytic probe for single lead ion monitoring. J Am Chem Soc 131(23):8221–8226. doi:10.1021/ja901132y

56. Xiang Y, Tong A, Lu Y (2009) Abasic site-containing DNAzyme and aptamer for label-free fluorescent detection of $Pb^{(2+)}$ and adenosine with high sensitivity, selectivity, and tunable dynamic range. J Am Chem Soc 131(42):15352–15357. doi:10.1021/ja905854a

57. Yin BC, Ye BC, Tan W, Wang H, Xie CC (2009) An allosteric dual-DNAzyme unimolecular probe for colorimetric detection of copper(II). J Am Chem Soc 131(41):14624–14625. doi:10.1021/ja9062426

58. Li T, Dong S, Wang E (2010) A lead(II)-driven DNA molecular device for turn-on fluorescence detection of lead(II) ion with high selectivity and sensitivity. J Am Chem Soc 132(38):13156–13157. doi:10.1021/ja105849m

59. Mor-Piperberg G, Tel-Vered R, Elbaz J, Willner I (2010) Nanoengineered electrically contacted enzymes on DNA scaffolds: functional assemblies for the selective analysis of Hg2+ ions. J Am Chem Soc 132(20):6878–6879. doi:10.1021/ja1006355

60. Dave N, Chan MY, Huang PJJ, Smith BD, Liu JW (2010) Regenerable DNA-functionalized hydrogels for ultrasensitive, instrument-free mercury(II) detection and removal in water. J Am Chem Soc 132(36):12668–12673. doi:10.1021/Ja106098j

61. Nie Z, Nijhuis CA, Gong J, Chen X, Kumachev A, Martinez AW, Narovlyansky M, Whitesides GM (2010) Electrochemical sensing in paper-based microfluidic devices. Lab Chip 10(4):477–483. doi:10.1039/b917150a

62. Lee SJ, Lee JE, Seo J, Jeong IY, Lee SS, Jung JH (2007) Optical sensor based on nanomaterial for the selective detection of toxic metal ions. Adv Funct Mater 17(17):3441–3446. doi:10.1002/adfm.200601202

63. Balaji T, El-Safty SA, Matsunaga H, Hanaoka T, Mizukami F (2006) Optical sensors based on nanostructured cage materials for the detection of toxic metal ions. Angew Chem 45(43):7202–7208. doi:10.1002/anie.200602453

64. El-Safty SA, Prabhakaran D, Ismail AA, Matsunaga H, Mizukami F (2007) Nanosensor design packages: a smart and compact development for metal ions sensing responses. Adv Funct Mater 17(18):3731–3745. doi:10.1002/adfm.200700447

65. El-Safty SA, Ismail AA, Matsunaga H, Hanaoka T, Mizukami F (2008) Optical nanoscale pool-on-surface design for control sensing recognition of multiple cations. Adv Funct Mater 18(11):1608–1608, 1485. doi:10.1002/adfm.200701059

66. Zhang J-T, Wang L, Luo J, Tikhonov A, Kornienko N, Asher SA (2011) 2-D Array photonic crystal sensing motif. J Am Chem Soc 133(24):9152–9155. doi:10.1021/ja201015c
67. Choi Y, Park Y, Kang T, Lee LP (2009) Selective and sensitive detection of metal ions by plasmonic resonance energy transfer-based nanospectroscopy. Nat Nanotechnol 4(11): 742–746. doi:10.1038/Nnano.2009.258
68. Wu C-S, Khaing Oo MK, Fan X (2010) Highly sensitive multiplexed heavy metal detection using quantum-dot-labeled DNAzymes. ACS Nano 4(10):5897–5904. doi:10.1021/nn1021988
69. Lee J, Jun H, Kim J (2009) Polydiacetylene-liposome microarrays for selective and sensitive mercury(ii) detection. Adv Mater 21(36):3674–3677. doi:10.1002/adma.200900639
70. Kagan D, Calvo-Marzal P, Balasubramanian S, Sattayasamitsathit S, Manesh KM, Flechsig G-U, Wang J (2009) Chemical sensing based on catalytic nanomotors: motion-based detection of trace silver. J Am Chem Soc 131(34):12082–12083. doi:10.1021/ja905142q
71. Zhou Y, Wang S, Zhang K, Jiang X (2008) Visual detection of copper(II) by azide- and alkyne-functionalized gold nanoparticles using click chemistry. Angew Chem 47(39): 7454–7456. doi:10.1002/anie.200802317
72. Zhang T, Cheng Z, Wang Y, Li Z, Wang C, Li Y, Fang Y (2010) Self-assembled 1-octadecanethiol monolayers on graphene for mercury detection. Nano Lett 10(11): 4738–4741. doi:10.1021/nl1032556
73. Sudibya HG, He Q, Zhang H, Chen P (2011) Electrical detection of metal ions using field-effect transistors based on micropatterned reduced graphene oxide films. ACS Nano 5(3): 1990–1994. doi:10.1021/nn103043v
74. Liu Y, Dong X, Chen P (2012) Biological and chemical sensors based on graphene materials. Chem Soc Rev 41(6):2283–2307. doi:10.1039/c1cs15270j
75. Kang Y, Walish JJ, Gorishnyy T, Thomas EL (2007) Broad-wavelength-range chemically tunable block-copolymer photonic gels. Nat Mater 6(12):957–960. doi:10.1038/nmat2032
76. Aguirre CI, Reguera E, Stein A (2010) Tunable colors in opals and inverse opal photonic crystals. Adv Funct Mater 20(16):2565–2578. doi:10.1002/adfm.201000143
77. Galisteo-Lopez JF, Ibisate M, Sapienza R, Froufe-Perez LS, Blanco A, Lopez C (2011) Self-assembled photonic structures. Adv Mater 23(1):30–69. doi:10.1002/adma.201000356
78. von Freymann G, Kitaev V, Lotsch BV, Ozin GA (2013) Bottom-up assembly of photonic crystals. Chem Soc Rev 42(7):2528–2554. doi:10.1039/c2cs35309a
79. Deng S, Yetisen AK, Jiang K, Butt H (2014) Computational modelling of a graphene Fresnel lens on different substrates. RSC Adv 4(57):30050–30058. doi:10.1039/C4ra03991b
80. Kong X-T, Butt H, Yetisen AK, Kangwanwatana C, Montelongo Y, Deng S, Da Cruz Vasconcellos F, Qasim MM, Wilkinson TD, Dai Q (2014) Enhanced reflection from inverse tapered nanocone arrays. Appl Phys Lett 105(5):053108. doi:10.1063/1.4892580
81. Martinez-Hurtado JL, Davidson CA, Blyth J, Lowe CR (2010) Holographic detection of hydrocarbon gases and other volatile organic compounds. Langmuir 26(19):15694–15699. doi:10.1021/la102693m
82. Naydenova I, Jallapuram R, Toal V, Martin S (2008) A visual indication of environmental humidity using a color changing hologram recorded in a self-developing photopolymer. Appl Phys Lett 92(3):031109. doi:10.1063/1.2837454
83. Vasconcellos FD, Yetisen AK, Montelongo Y, Butt H, Grigore A, Davidson CAB, Blyth J, Monteiro MJ, Wilkinson TD, Lowe CR (2014) Printable surface holograms via laser ablation. ACS Photonics 1(6):489–495. doi:10.1021/Ph400149m
84. Yetisen AK, Jiang L, Cooper JR, Qin Y, Palanivelu R, Zohar Y (2011) A microsystem-based assay for studying pollen tube guidance in plant reproduction. J Micromech Microeng 21 (5):054018. doi:10.1088/0960-1317/21/5/054018
85. Yetisen AK, Naydenova I, Vasconcellos FC, Blyth J, Lowe CR (2014) Holographic sensors: three-dimensional analyte-sensitive nanostructures and their applications. Chem Rev 114 (20):10654–10696. doi:10.1021/cr500116a

86. Yetisen AK, Butt H, da Cruz Vasconcellos F, Montelongo Y, Davidson CAB, Blyth J, Chan L, Carmody JB, Vignolini S, Steiner U, Baumberg JJ, Wilkinson TD, Lowe CR (2014) light-directed writing of chemically tunable narrow-band holographic sensors. Adv Opt Mater 2(3): 250–254. doi:10.1002/adom.201300375

87. Yetisen AK, Montelongo Y, da Cruz Vasconcellos F, Martinez-Hurtado JL, Neupane S, Butt H, Qasim MM, Blyth J, Burling K, Carmody JB, Evans M, Wilkinson TD, Kubota LT, Monteiro MJ, Lowe CR (2014) Reusable, robust, and accurate laser-generated photonic nanosensor. Nano Lett 14(6):3587–3593. doi:10.1021/nl5012504

88. Tsangarides CP, Yetisen AK, da Cruz Vasconcellos F, Montelongo Y, Qasim MM, Wilkinson TD, Lowe CR, Butt H (2014) Computational modelling and characterisation of nanoparticle-based tuneable photonic crystal sensors. RSC Adv 4(21):10454–10461. doi:10.1039/C3RA47984F

89. Yetisen AK, Montelongo Y, Qasim MM, Butt H, Wilkinson TD, Monteiro MJ, Lowe CR, Yun SH (2014) Nanocrystal bragg grating sensor for colorimetric detection of metal ions. (under review)

90. Akram MS, Daly R, Vasconcellos FC, Yetisen AK, Hutchings I, Hall EAH (2015) Applications of paper-based diagnostics. In: Castillo-Leon J, Svendsen WE (eds) Lab-on-a-chip devices and micro-total analysis systems. Springer, New york

91. Yetisen AK, Akram MS, Lowe CR (2013) Paper-based microfluidic point-of-care diagnostic devices. Lab Chip 13(12):2210–2251. doi:10.1039/c3lc50169h

92. Yetisen AK, Volpatti LR (2014) Patent protection and licensing in microfluidics. Lab Chip 14 (13):2217–2225. doi:10.1039/c4lc00399c

93. Farandos NM, Yetisen AK, Monteiro MJ, Lowe CR, Yun SH (2014) Contact lens sensors in ocular diagnostics. Adv Healthc Mater. doi:10.1002/adhm.201400504

94. Yetisen AK, Martinez-Hurtado JL, Garcia-Melendrez A, Vasconcellos FC, Lowe CR (2014) A smartphone algorithm with inter-phone repeatability for the analysis of colorimetric tests. Sens Actuators, B 196:156–160. doi:10.1016/j.snb.2014.01.077

95. Yetisen AK, Martinez-Hurtado JL, da Cruz Vasconcellos F, Simsekler MC, Akram, Lowe CR (2014) The regulation of mobile medical applications. Lab Chip 14(5):833–840. doi:10.1039/c3lc51235e

Chapter 5
Holographic Glucose Sensors

Rapid glucose sensors have applications in the screening, diagnosis and monitoring of diabetes at point of care. This chapter demonstrates the design, fabrication and clinical trial of reusable holographic glucose sensors. Holographic sensors comprised boronic acid derivative functionalised acrylamide matrices, which consisted of Bragg diffraction gratings that colorimetrically report on the concentration of glucose in aqueous solutions [1]. The optical properties of the sensor were designed and characterised by computational analysis. The sensors were fabricated by combining the advantages of multi-beam interference and in situ size reduction of silver metal (Ag^0) nanoparticles (NPs) by single-pulse laser writing [2]. Fully-quantitative narrow-band (monochromatic) readouts were attained through spectrophotometry. The advantages of holographic sensors over other sensing mechanisms are (i) reusability, (ii) amenable to mass manufacturing through laser writing, (iii) readouts in visible as well as near-infrared regions of the spectrum, and (v) reproducibility to sense glucose concentrations up to 400 mM using a low sample volume (<500 μl). Interference due to other metabolites such as lactate and fructose was also evaluated. Trials of the sensor in the urine samples of diabetic patients demonstrated that the sensor had improved performance as compared to Multistix® 10 SG read by CLINITEK Status®, while having comparable performance with fully-automated Dimension® Clinical Chemistry System. Holographic glucose sensors may have clinical applicability for diabetes screening or diagnosis of bacterial urinary tract infections.

5.1 Diabetes Mellitus

Diabetes is one of the most challenging health problems of the 21st century. The global prevalence of diabetes has increased dramatically, and is now a worldwide public health concern [3, 4]. Currently, 382 million people live with diabetes. This epidemic on the rise all over the world, and it has overwhelmed the healthcare systems. The number of people with diabetes is estimated to reach 592 million in less than 25 years [5]. In 2035, one in ten people is likely to have diabetes. Diabetes

© Springer International Publishing Switzerland 2015
A.K. Yetisen, *Holographic Sensors*, Springer Theses,
DOI 10.1007/978-3-319-13584-7_5

has been known to be 'a disease of the wealthy'. However, today 80 % of people with diabetes live in low- and middle-income countries, and the socially disadvantaged in any country are the most vulnerable to the disease. The number of people with diabetes is rapidly increasing in the Middle East, Western Pacific, sub-Saharan Africa and South-East Asia, where economic development has transformed lifestyles. These rapid transitions have resulted in high rates of obesity and diabetes; developing countries are facing a healthcare challenge coupled with inadequate resources to protect their population. According to International Diabetes Federation, in sub-Saharan Africa, where resources lack and governments may not prioritise screening for diabetes, this proportion is as high as 90 % in some countries [5]. The new estimates show an increasing trend towards younger people developing diabetes [5]. The financial burden of diabetes is annually 548 billion dollars, which is 11 % of the global healthcare expenditure [5]. Yet, it is estimated that 175 million people are undiagnosed today [5]. This is because there are few symptoms during the early years of type 2 diabetes, or those symptoms are not recognised as being related to diabetes. Type 2 diabetes can go unnoticed and undiagnosed for years. In such cases, those affected are unaware of the long-term damage being caused by diabetes.

Diabetes can be screened with blood glucose meters and urine strip tests. These tests are low cost, but considering that 1 billion people live on less than $1.25 a day, they are not affordable [6]. For example, glucometers ($40), a lancing device ($15), lancets (10¢) and test strips (50¢) can cost up to ~ $100 for testing hundred people in a developing world country. On the other hand, the urine dipstick test, which costs about 50¢, has low sensitivity and often provides erroneous results due to subjective reading. The development of diagnostic devices that are low cost, reusable, user friendly, non-invasive and reliable will help deprived communities. Once diagnosed, the management of diabetes becomes a liability for diabetics. Most diabetics exhibit hypo/hyperglycemia requiring tight control of blood glucose concentration. In diabetes management, diabetics need to regularly measure their blood glucose by finger pricking up to five times a day to control the concentration of glucose in blood, which requires at least three insulin injections daily. This invasive practice reduces the rate of patient compliance and results in less effective glycaemic control. In particular, young adults have lower compliance as compared to older patients in providing samples to healthcare workers and self-managing their diabetes. Measuring the concentration of glucose in urine is a less accurate way of estimating the concentration of glucose in blood since urine tested is produced several hours before the test. Currently, there is an ever-increasing need for low-cost sensors to non-invasively manage diabetes. An ideal glucose sensor should operate under physiological pH, ionic strength (IS) and in the presence of biological fluids, and measure glucose concentrations with high accuracy both within and outside the normal blood glucose concentration range (4.2–6.7 mM) [7] with a quick turnaround time (<5 min). Development of non-invasive and accurate diagnostics that are easily manufactured, robust, and reusable will provide monitoring high-risk individuals in any clinical or point-of-care environment, particularly in the developing world.

5.2 Holographic Glucose Sensors

Optical sensors consisting of polyacrylamide (pAAm) matrices have been functionalised with boronic acid derivatives. 3-(Acrylamido)phenylboronic acid (3-APB) forms a reversible covalent bond with *cis* diol [8]. Boronic acid ($pK_a = \sim 8.8$) at low pH values is in an uncharged and trigonal planar configuration (1); while at higher pH values (pH > pK_a), the trigonal form reacts with OH⁻ to form the more stable negatively charged tetrahedral state (2), which can bind to *cis* diol groups more readily (3–4) (Fig. 5.1a) [9, 10]. The binding of *cis* diols forms boronate anions, and subsequently swells the hydrogel through a Donnan osmotic pressure increase [11]. This swelling increases the lattice spacing, which shifts the Bragg peak to longer wavelengths (Fig. 5.1b–d).

Holographic sensors comprising of 3-APB have been fabricated and their sensitivity was optimised for glucose detection [12]. Maximal sensitivity was measured at a 3-APB concentration of ~ 20 mol% [8, 13]. The hologram displayed a red Bragg

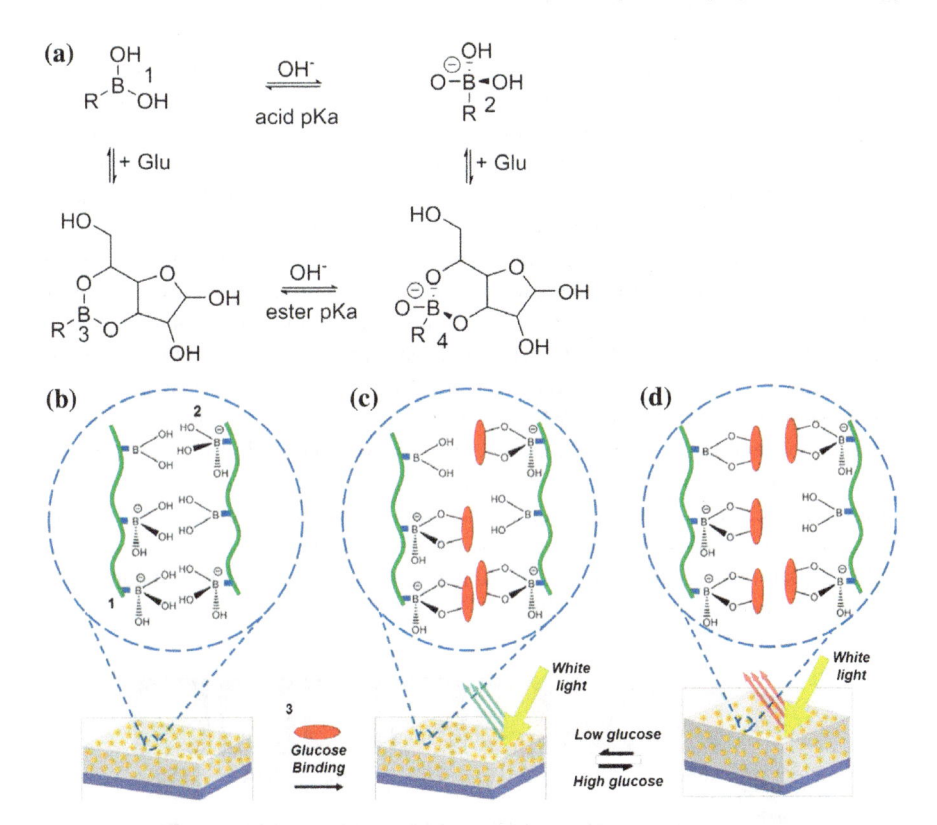

Fig. 5.1 Principle of operation of a Bragg grating-based glucose sensor. **a** Simplified illustration of the complexation equilibrium between the boronic acid derivatives and glucose. Glu = Glucose, **b** Reversible swelling of the sensor by glucose modulates both the Ag^0 NP distribution spacing and the refractive-index contrast, and systematically shifts the Bragg peak from **c** shorter to **d** longer wavelengths as the hydrogel expands in the direction normal to the underlying substrate

peak shift as a function of glucose concentration (2–10 mM) across the physiological range. When the glucose molecules perfuse into the polymer matrix, the pK_a of the boronic acid-glucose complex systematically decreases through the stabilisation of the charged tetrahedral phenylboronate anion. As a result of producing charged groups, the Donnan osmotic pressure of the polymer matrix increases, leading to the hologram to imbibe more water. Therefore, the hologram swells and its Bragg peak shifts to longer wavelengths. The glucose-*cis* diol binding is reversible due to the formation of the covalent bond in aqueous media. When the hologram is rinsed with glucose-free medium, the hologram contracts and returns to its original Bragg peak position [13]. Such a capability allows the sensor to be used in continuous sensing of dynamic changes in glucose concentration. However, 3-APB also binds to other *cis* diol containing species such as fructose and lactate in the biological fluids [14]. For example, the presence of fructose or lactate in blood and its interaction with the phenylboronic acid may lead to an artificially elevated glucose readout. An additional limitation of hydrogel-based holographic sensors is the dependency on IS; solutions with higher IS contract the matrix [15]. In order to overcome these challenges, holographic sensors incorporating new boronic acid derivatives that bind to glucose at physiological pH conditions have been developed. For example, 2-acrylamidophenylboronate (2-APB) was synthesised and its principle to binding glucose was investigated [16, 17]. 2-APB predominantly adopts a zwitterionic tetrahedral form at physiological pH values and it has a tendency to complex with glucose rather than lactate. The binding mechanism of 2-APB to glucose was also unaffected by the pH variation within the physiological range. Other studies investigated strategies to reduce the fructose interference. For example, the incorporation of tertiary amine monomer or quaternary monomer such as 3-(acrylamidopropyl)trimethylammonium chloride (ATMA) into holographic sensors containing phenylboronic acids improved the selectivity for glucose [18, 19]. This synthesis approach was used to quantify the concentration of glucose in blood in vitro [20]. This study involved a hologram comprising of 3-APB, and ATMA to measure human blood plasma samples at glucose concentrations of 3–33 mmol/L over an extended period for application in continuous monitoring (Fig. 5.2). The sensor had a performance comparable to an electrochemical sensor. The measurement accuracy was not affected in the presence of common antibiotics, diabetic drugs, pain killers and endogenous substances [21]. Table 5.1 summarizes the tested holographic sensors, their experimental conditions, and their respective wavelength shifts.

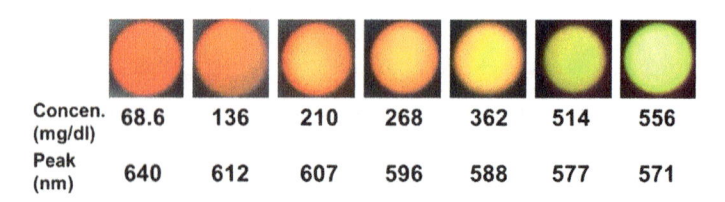

| Concen. (mg/dl) | 68.6 | 136 | 210 | 268 | 362 | 514 | 556 |
| Peak (nm) | 640 | 612 | 607 | 596 | 588 | 577 | 571 |

Fig. 5.2 A holographic glucose sensor in serum samples. The holographic matrix comprising of MBA (3 mol%), 3-APB (12 mol%), ATMA (12 mol%) in human blood plasma as a function of glucose concentration (pH 7.4 and 37 °C). Reprinted with permission from [20] Copyright 2007 American Association for Clinical Chemistry

The need for minimally invasive, easy to use glucose sensors has motivated investigation of ophthalmic glucose sensors, allowing the detection of glucose in ocular fluid such as tears [22]. The feasibility of non-invasive monitoring of the concentration of glucose was investigated by incorporating a 3-APB-based holographic sensor into a contact lens [23, 24]. Another experiment was conducted through implanting the hologram subcutaneously just below the eye of a rabbit, followed by xylazine-containing anaesthetic, which increased the concentration of glucose in blood [25]. In ~ 3 min, the Bragg peak shifted by 25 nm, which was correlated with simultaneous measurement of glucose (~ 10 mg%) in the blood. Such a sensor can also be fabricated to semi-quantitatively measure the concentration of glucose by displaying different colours or images [26]. However,

Table 5.1 The compositions of holographic glucose sensors and their Bragg peak shifts

Ligand (mol%)	pH	Concentration (mM) [Bragg peak shift (nm)]	References
3-APB (12)	7.4	2 (20), 11 (70)	[13]
3-APB (20)	7.4	2 (50), 11 (220)	[13]
3-APB (25)	7.4	2 (40), 7.4 (120)	[8]
3-APB (12)	7.4	2 (20), 11 (70)	[8]
3-APB (20)	7.4	2 (60), 11 (220)	[8]
5-F 2MAPB (15)	7.4	2 (30), 11 (120)	[8]
2-APB (20)	7.1	2 (0), 12 (0)	[17]
2-APB (20)	5.8	2 (−5), 12 (−23)	[17]
2-APB (20)	6.5	2 (−5), 12 (−23)	[17]
2-APB (20)	7.0	2 (−5), 12 (−23)	[17]
2-APB (20)	7.8	2 (−5), 12 (−23)	[17]
3-APB (12)	5.8	2 (−3), 12 (−6)	[17]
3-APB (12)	6.5	2 (−5), 12 (−12)	[17]
3-APB (12)	7.0	2 (−15), 12 (−35)	[17]
3-APB (12)	7.4	2 (−17), 12 (−45)	[17]
3-APB (12)	7.8	2 (−10), 12 (−25)	[17]
3-APB (12)	7.4	2 (20), 9 (70)	[18]
3-APB (12) + DAPA (16)	7.4	2 (−30), 11 (−60)	[18]
3-APB (12) + ATMA (12)[a]	7.4	4.9 (−60), 8.9 (−70)	[20]
2-APB (20) + ATMA (3)	7.4	2 (8), 9 (30)	[28]
2-APB (20) + PEG (3)	7.4	2 (6), 9 (20)	[28]
2-APB (20) + AETA (3)	7.4	2 (4), 9 (17)	[28]
2-APB (20)	7.4	2 (3), 9 (15)	[28]
2-APB (20) + ATMA (9)	6.5	2 (−4), 11 (−7)	[28]
2-APB (20) + ATMA (9)	7.0	2 (−4), 11 (−7)	[28]
2-APB (20) + ATMA (9)	7.4	2 (−4), 11 (−7)	[28]
2-APB (20) + ATMA (9)	7.8	2 (−4), 11 (−7)	[28]

IS of 150 mM at 30–37 °C
[a] blood glucose

controversy still remains whether the concentration of glucose in tear fluid and blood is correlated [27]. Tear fluid composition is subject to change based on the method of sample collection. Erroneous sample collection methods have caused disagreement about the concentrations of analytes in tear fluid [22].

5.3 Computational Modelling of Holographic Glucose Sensors

Computation of the interaction between a Bragg grating and an electromagnetic field using finite-difference time domain algorithms as the numerical analysis provided a theoretical model of the optical properties for the holographic sensor. The simulated geometry consisted of a computed photonic structure, which had superpositioned laser beam interference inside a theoretical Ag^0 NP impregnated pAAm matrix, created by three beams: (1) incident beam (reference), (2) beam reflected from the mirror (object) and (3) beam reflected internally at the pAAm matrix-water interface. Assuming that the photonic structure only consisted of a multilayer diffraction grating with a lattice spacing of $\Lambda/2$, the optical properties were defined based on angular-resolved measurements and effective index of refraction of the pAAm matrix. Figure 5.3 shows the computed spectral response as a laser beam (532 nm) based on the wave equation was propagated onto the

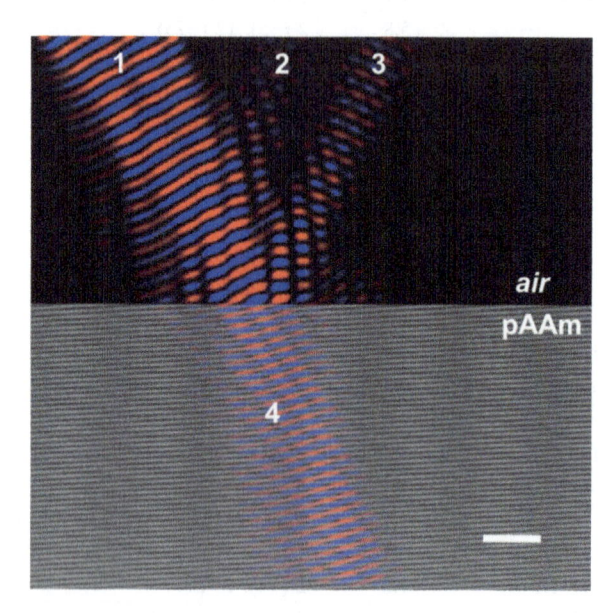

Fig. 5.3 Simulated optical response of patterned Ag^0 NP impregnated pAAm matrix. A coherent source *1* of an incident wave ($\lambda = 532$ nm) originating from air is propagated onto the sensor. The photonic structure reflects *3*, refracts *4* and diffracts *2*, which allows offsetting the diffraction for reporting on the concentrations of analytes. The colours represent the phase of the electromagnetic field. *Scale bar:* 2 µm. Reprinted with permission from [29] Copyright 2014 The American Chemical Society

photonic structure along the pAAm matrix's transversal section confined within a 20×10 μm^2 area. The colours from blue to red represent the phase of the electromagnetic field. Fresnel's law describes the paths of the incoming **1**, reflected **3** and the refracted waves of light **4**. The diffracted wave **2** observed at angles away from specular reflection, obeying Bragg's law, $\lambda_{peak} = 2\Lambda\sin(\theta)$, where λ_{peak} is the wavelength of the first order diffracted light at the maximum intensity *in vacuo*, n_0 is the effective index of refraction of the recording medium, Λ is the spacing between the two consecutive recorded NP layers, and θ is the Bragg angle. The offset diffraction from the device produced a narrow-band readout signal, allowing the differentiation of small changes in optical properties of the device due to different types and concentrations of analyte. These simulations allow designing holographic sensors with predictive optical properties. For example, simulations can be used to estimate the angle of diffraction or the diffraction efficiency required for a specific application.

5.4 Fabrication of Holographic Glucose Sensors

The monomer mixture (5 mmol) consisted of acrylamide (AAm), *N,N'*-methylenebisacrylamide (MBAAm) and 3-APB (Table 5.2). The solution was mixed (1:1, v/v) with DMPA in DMSO (2 %, w/v). The mixture was copolymerised on an O$_2$-plasma-treated PMMA substrate (Fig. 5.4a). Ag$^+$ ions were perfused into poly (AAm-co-3-APB) matrix, and they were reduced to Ag^0 NPs in situ using a photographic developer (Fig. 5.4b). A holographic sensor was formed by photochemically patterning the matrix via a single 6 ns Nd:YAG laser pulse ($\lambda = 532$ nm,

Table 5.2 Composition of glucose-responsive hydrogels

Monomer	Molecular weight (g/mol)	Molarity (%)
AAm	71.08	75–80
MBAAm	154.17	1.5
3-APB	190.99	10–20
DMPA	256.30	2 (wt%) in DMSO

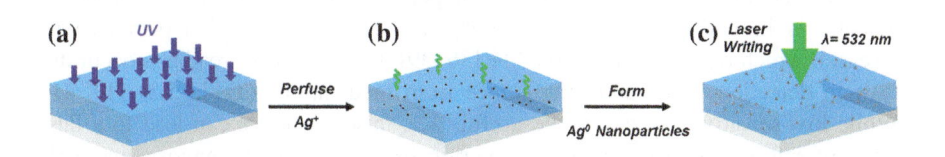

Fig. 5.4 Fabrication of holographic glucose sensors. **a** Free radical copolymerisation of AAm, MBAAm and 3-APB, **b** perfusion of Ag$^+$ ions into the poly(AAm-co-3-APB) matrix and formation of Ag^0 NPs using a photographic developer, and **c** formation of diffraction gratings in Ag^0 NP impregnated matrix

350 mJ), directed at the sample elevated at 5° from the surface plane and backed by a mirror (object) (Fig. 5.4c).

Analysis of the NP size distribution, angular measurements and determination of apparent pK_a value of the hydrogel matrix in combination with computational modelling of optical properties allowed optimisation and utilisation of the diffraction properties for sensing applications. TEM imaging was used to estimate the NP size distribution in the matrix. The holographic glucose sensor had a diffraction grating consisting of Ag^0 NPs ($\sim\varnothing$ 15–25 nm) organised by the multi-beam interference of a single 6 *ns* laser (λ = 532 nm, 200 mJ) pulse within a $\sim$10 μm thick pAAm matrix functionalised with 3-APB. Figure 5.5a, b illustrates the cross sectional images of Ag^0 NP impregnated poly(AAm-co-3-APB) matrix before and after photochemical patterning, respectively. The Ag^0 NPs after single-pulse laser-induced photochemical patterning reduced in size as illustrated by the TEM images of the transverse plane of the matrix. Figure 5.5c, d shows the NP thresholds set for size distribution analysis for Fig. 5.5a, b, respectively. Figure 5.5e–i illustrates the matrix before and after patterning, respectively. Figure 5.6 shows the distribution of Ag^0 NPs across the cross section of the matrix. The images of the Ag^0 NPs in situ showed the reduction of size from $\varnothing$ 17 ± 11 nm (n = 83) to $\varnothing$ 10 ± 9 nm (n = 221) before and after patterning, respectively. The changes in particle size, density and periodicity of the Ag^0 NPs defined the sensor's spectral response to the intensity and diffraction at different wavelengths.

5.5 Holographic Glucose Sensors for Urinalysis

The measurement of urine glucose has diagnostic applicability in a number of clinically relevant conditions [30]. Under normal conditions, the excretion rate of glucose in urine ranges between 0.30 and 1.70 mmol/24 h [7]. Since most filtered glucose is normally reabsorbed, an elevated urine glucose concentration indicates either impaired tubular reabsorption of glucose (e.g. familial renal glycosuria), or more commonly, hyperglycemia that exceeds the kidney's reabsorptive capacity (e.g. diabetes mellitus) [31, 32]. Conversely, a low concentration of urine glucose may be found in urinary tract infections due to the bacterial metabolism of glucose [33]. Existing colorimetric and electrochemical tests are based on the glucose oxidase reaction [34, 35]. However, their performance in detecting undiagnosed diabetes is limited due to low sensitivity (i.e. correctly identified patients with disease), which ranges from 21 to 64 % [36–39]. False negative readings occur due to high detection limits and interference from medications (Table 5.3) [40, 41]. While low-sensitivity tests may be useful [42], false negatives can lead to a false sense of safety among users, and more critically delay correct diagnosis and early treatment [38, 43].

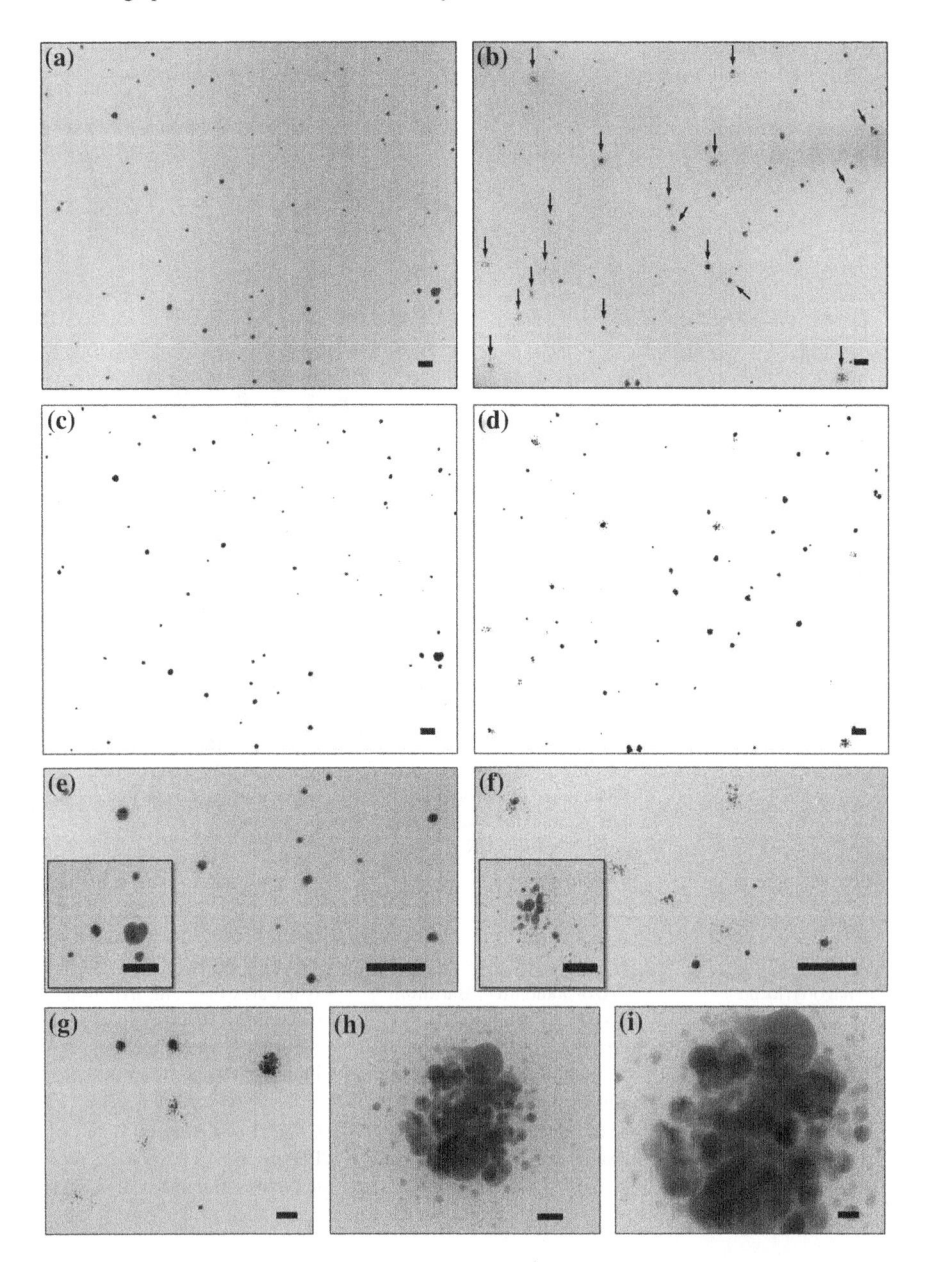

Fig. 5.5 TEM images of the transversal section of Ag^0 NP impregnated poly(AAm-co-3-APB) matrix **a, e** before and **b, f–i** after exposure to the laser pulse showing reduction in the diameter of Ag^0 NPs at patterned planes. The NPs with reduced diameter (ø) are indicated by the *arrows* in **b. c, d** illustrates the NP thresholds set for size distribution analysis for **a, b**. *Scale bars* **a, b, c, d, g** = 100 nm, **e, f** = 200 nm, **h** = 20 nm, **i** = 10 nm. Reprinted with permission from [29] Copyright 2014 The American Chemical Society

Fig. 5.6 Size distribution of Ag^0 NPs before and after laser-induced photochemical patterning in the poly(AAm-co-3-APB) matrix

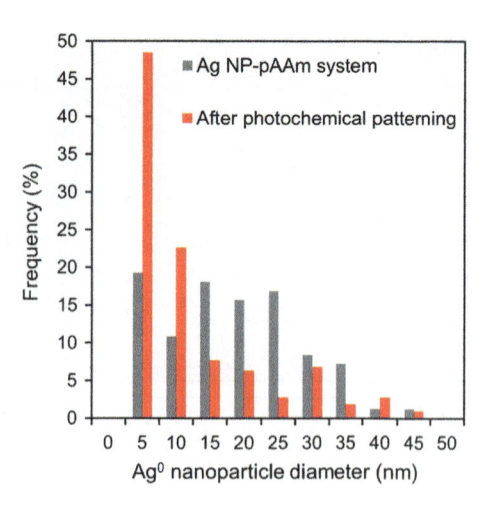

Table 5.3 Inhibition of glucose oxidase due to various drugs and chemical agents [41, 44]

Chemical	Brand name	Use
Mecetronium ethylsulphate	Sterillium	Disinfectant
Sodium 2-mercaptoethanesulphonate	Uromitexan and Mesnex	An adjuvant in cancer chemotherapy involving cyclophosphamide and ifosfamide
Nitrofurantoin	Niftran, Furadantin, Furabid, Macrobid, Macrodantin, Nitrofur Mac, Nitro Macro, Nifty-SR and Martifur-MR	Antibiotic used in treating urinary tract infections caused by *E. coli*
Acetylsalicylic acid	Aspirin	An analgesic to relieve minor aches and pains, as an antipyretic to reduce fever, and as an anti-inflammatory medication
Phenazopyridine hydrochloride	Azo-Standard®, Baridium®, Nefrecil®, Prodium®, Pyridate®, Pyridium®, Sedural®, Uricalm®, Uristat®, Uropyrine® and Urodine®	Alleviates the pain, irritation, discomfort, or urgency caused by urinary tract infections, or injury to the urinary tract
Levodopa	Sinemet, Parcopa, Atamet, Stalevo, Madopar and Prolopa	Clinical treatment of Parkinson's disease and dopamine-responsive dystonia

5.5.1 Holographic Glucose Sensor Readouts

The response of the holographic sensors was first tested in phosphate buffers. As the concentration of glucose increased from 0.1 to 10.0 mM under physiological conditions (pH 7.4, IS = 150 mM), the Bragg peak originating from the structure

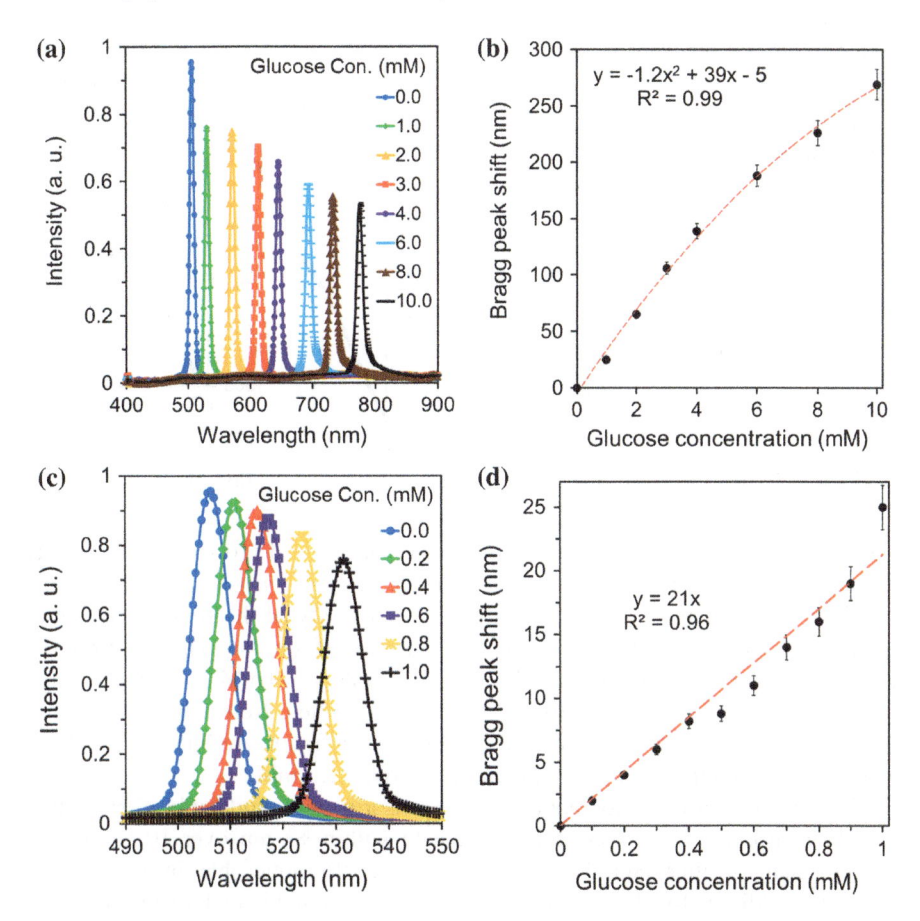

Fig. 5.7 Readouts of a holographic sensor due to variation in the concentration of glucose. **a** Diffraction spectra of a holographic sensor swollen by different glucose concentrations in phosphate buffers (pH = 7.4, IS = 150 mM, 24 °C). The largest Bragg peak shift is at 775 nm (10.0 mM glucose) and the smallest is at 505 nm (glucose-free). **b** The Bragg peak shift due to 10.0 mM glucose as a function of glucose concentration over three trials, **c** Bragg peak shifts of a holographic sensor swollen by contact with different glucose concentrations <1 mM over the physiological range, **d** The Bragg peak shifts over three trials as a function of glucose concentration <1.0 mM. Standard *error bars* represent three independent glucose samples

systematically shifted from 505 to 775 nm (Fig. 5.7a). As the hologram expanded normal to the underlying substrate, the diffraction efficiency of the peaks decreased. This behaviour can be attributed to the decrease in the density of Ag^0 NPs present in the periodic regions of the hologram, which reduces the effective index contrast between the patterned regions and the poly(AAm-co-3-APB) matrix. The Bragg peak shift over three trials as a function of glucose concentration is shown in Fig. 5.7b. The sensing mechanism was reversible; the Bragg peak shifted to shorter wavelengths as the glucose concentration was decreased. During the readouts, the sample solutions were kept at 24 °C, and agitated using a magnetic stirring bar

Fig. 5.8 The response time
of the holographic glucose
sensor. When 1.0 and
10.0 mM glucose were added
into the reservoir, the sensor
equilibrated in ∼1 h at 24 °C

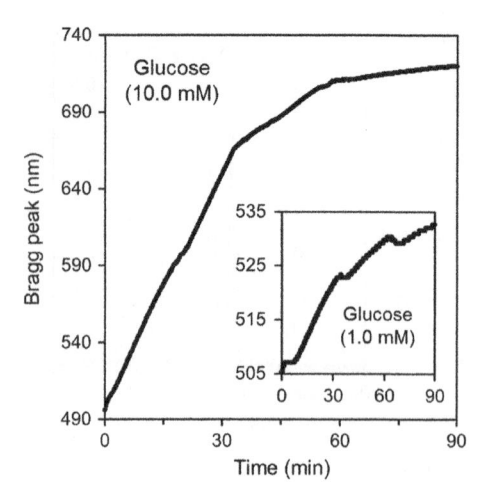

(micro flea, 2 mm). The sensor was reset using acetic acid (10 mol%, v/v). Other
acids that will decrease the pH can also be used. The same sensor may therefore be
used for repeat analyses. To demonstrate the sensitivity of the sensor, concentrations <1.0 mM were analysed. As the concentration of glucose increased from 0.0
to 1.0 mM under physiological conditions (pH 7.4, IS = 150 mM, 24 °C), the Bragg
peak of the holographic sensor systematically shifted from 507 to 532 nm
(Fig. 5.7c). The Bragg peak shift and glucose concentration are linearly correlated
(Fig. 5.7d), allowing the determination of glucose <1.0 mM.

The readouts of the holographic glucose sensor were reproducible. Consecutive
swelling/shrinking steps were reproducible to within ±3 nm over 10 successive
buffer changes after equilibrium was reached. No noticeable hysteresis was
detected. Additionally, the sensor's response was measured in real time for both 1.0
and 10.0 mM glucose (Fig. 5.8). Measurements at low glucose concentrations
(<1 mM) required ∼70 min to reach ∼90 % equilibrium, whereas measurements at
higher glucose concentrations (>1 mM) required shorter times (∼50 min) to reach
∼90 % equilibrium. The Bragg peak returned to the original position when PBS
was added and equilibrated in ∼1 h. The apparent pK_a value of the poly(AAm-co-
3-APB) matrix was ∼8.5 based on the Henderson-Hasselbalch equation, while the
sensor displayed improved sensitivity as the pH was increased from 7.0 to 9.5
(Fig. 5.9, see inset for the colorimetric readouts).

5.5.2 Holographic Glucose Sensor Readouts in Artificial Urine

Artificial urine solutions were prepared as reported previously [45]. Two stock
solutions (500 mL) of artificial urine were prepared with and without glucose
(10.0 mM). The solutions were filtered (pore size = 0.22 μm). The pH was adjusted

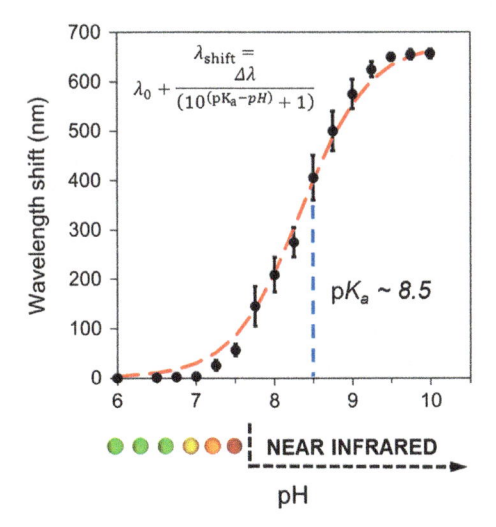

Fig. 5.9 Determination of apparent pK_a value of the poly(AAm-co-3-APB) matrix. The pK_a value was calculated using the Henderson-Hasselbalch equation, where $\lambda_{shift,\ 0\text{-}20}$ = step Bragg peak shift, λ_0 = initial wavelength, and $\Delta\lambda = (\lambda_{max} - \lambda_0)$ overall Bragg peak shift difference (n = 48). The *inset* shows colorimetric readouts of the sensor as a function of pH. Standard *error bars* represent three independent samples. Reprinted with permission from [29] Copyright 2014 The American Chemical Society

using HCl (1 M) and NaOH (1 M). The glucose-containing solution was serially diluted with the glucose-free solution to obtain standard concentrations of 0.1–10.0 mmol/L glucose at a constant IS ($\sim$210 mmol/L). Artificial urine solutions containing lactate and fructose were prepared using the same protocol in the concentrations of 0.2–10.0 mM and 0.1–2.0 mM, respectively. The artificial urine solutions were immediately used and fresh solutions were prepared for each trial. The sensor's response to an increase in clinically relevant concentrations of glucose (1.0–10.0 mM) in artificial urine showed a red Bragg shift under physiological conditions (pH 7.40). Figure 5.10a illustrates a typical narrow-band spectral readout at pH 7.40, indicating a decrease in the diffraction efficiency and peak broadening exhibited by the photonic structure as the concentration of glucose increased in artificial urine.

The trend of the readout was asymptotical due to an inverse correlation between the concentration of Ag^0 NPs and their lattice spacing (Λ) in the poly(AAm-co-3-APB) matrix. Hence, the maximum peak intensity (I_{max}) and its corresponding wavelength λ_{max} was inversely correlated to the periodicity of the Ag^0 NP regions because of the reduction in the contrast of the effective refractive index during lattice expansion. This relation can be expressed as $I_{max} = c/\lambda_{max}$, where c is a constant, assuming that asymptotes approach zero. However, in the poly

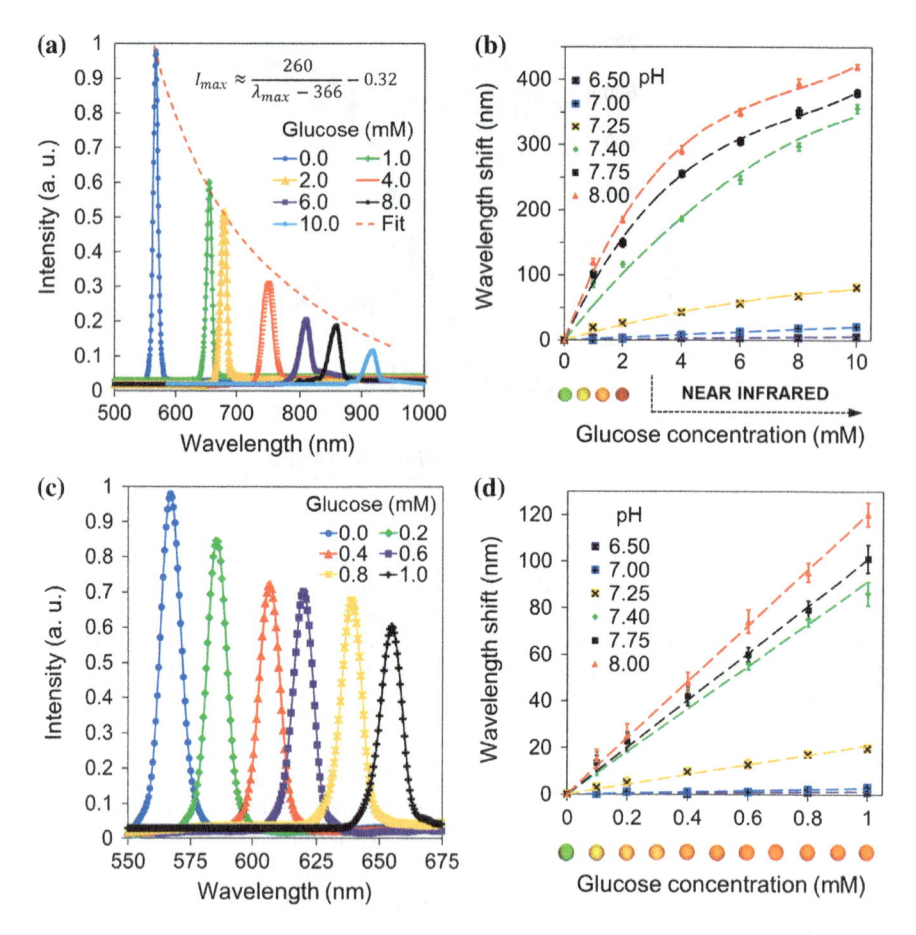

Fig. 5.10 The response the holographic sensor (20 mol% 3-APB) to variation in the concentration of glucose below 1.0 and 10.0 mM in artificial urine at 24 °C. **a** An increase in the glucose concentration of artificial urine solutions (pH 7.40, 24 °C) swelled the poly(AAm-co-3-APB) matrix, thus red shifting the Bragg peak, while also showing a correlation between the intensity and the wavelength measured (n = 7). Constants for the fit: $c = 260$ nm, $\lambda_0 = 366$ nm, $I_0 = 0.32$, and a.u. = arbitrary units. **b** Spectral readout at pH values from 6.50–8.00 (n = 126). The *inset* shows the colorimetric response at pH 7.40. **c** A typical sensor response to variation in glucose concentrations <1.0 mM. **d** The Bragg peak shift as a function of glucose concentration <1.0 mM shows reproducibility ($\pm \sim 5$ nm) over three trials. The *inset* shows the colorimetric response at pH 7.40. Standard *error bars* represent three independent samples. Reprinted with permission from [29] Copyright 2014 The American Chemical Society

(AAm-co-3-APB) matrix, where the expansion was finite and the asymptotes did not approach zero; the maximum intensity (I_{max}) at a given λ_{max}:

$$I_{max} \approx I_0 + \frac{c}{\lambda_{max} - \lambda_0} \tag{5.1}$$

where I_0 and λ_0 represent the asymptotes of the curve and λ_{max} is $2n\Lambda\cos(\theta)$, obeying Bragg's law, in which n is the effective index of refraction, and θ is the angle of illumination from the normal. Additionally, the displacement of the asymptotes regarding the first order could be attributed to other factors such as the scattering strength of each Ag^0 NP, which increased at Mie plasmon resonances in the blue/green region, hence the total amount of scattering decreased as the Bragg resonance shifted to longer wavelengths. The Bragg peak was ~ 565 nm for glucose-free artificial urine, and additions of up to 10.0 mM glucose shifted this peak systematically by 21, 81, 356, 379 and 420 nm, at pH values of 7.00, 7.25, 7.40, 7.75 and 8.00 (Fig. 5.10b); with the limit of detections of 0.61, 0.50, 0.41, 0.36, 0.26 mM, respectively. The limit of detection represents three times average standard deviation divided by the slope. The diffraction exhibited green, yellow, orange and red light before moving into the near-infrared region with further increases in glucose concentration (inset in Fig. 5.10b). Therefore, at an apparent pK_a of 8.5, the glucose bound with the tetrahedral form with degrees of ionisation of 0.4, 3.2, 4.9, 17.8 and 25.7 % at pH values of 7.00, 7.25, 7.40, 7.75 and 8.00 and subsequently reached equilibrium. At low concentrations, glucose bound with the boronic acid groups in a tetrahedral coordination form, in which the binding transformed, although kinetically slower, to a trigonal planar form at higher concentrations [46]. A decrease in the slope was consequently observed at the higher concentration range. Another explanation for the change in the slope was that as the hydrogel matrix expanded, the rate of swelling slowed down due to a decrease in the elasticity. The potential clinical utility of the sensor in detecting hypoglycosuria associated with urinary tract infections was tested by quantifying glucose concentrations below 1.0 mM. When bacteria are present in urine, they metabolise existing glucose, decreasing its concentration below 1.0 mM [47]. Therefore, measuring low concentrations of urine glucose can be used as a surrogate for rapid screening of urinary tract infections [33]. The detection and early treatment of urinary tract infections may reduce the risk of chronic kidney failure due to renal scarring [48, 49]. Figure 5.10c shows quantification of the glucose concentration from 0.0 to 1.0 mM in artificial urine solutions (pH 7.40) by the systematic shift of the Bragg peak to longer wavelengths, displaying a typical narrow-band spectral readout. Figure 5.10d illustrates quantification of the glucose concentration from 0.0 to 1.0 mM in artificial urine solutions at different pH values (see inset in Fig. 5.10d for colorimetric response). A Bragg peak shift from 3 to 120 nm with a systematic increase in pH from 7.00 to 8.00 provided concentration detection limits ranging from 240 to 90 μM, respectively.

5.5.3 Lactate and Fructose Interference

The *cis* diol groups of 3-APB can competitively bind to lactate and fructose. Lactate is present at low concentrations in urine 0.00–0.25 mmol/L [50] and it may increase during physical activity. Through its α-hydroxy acid, lactate can competitively bind

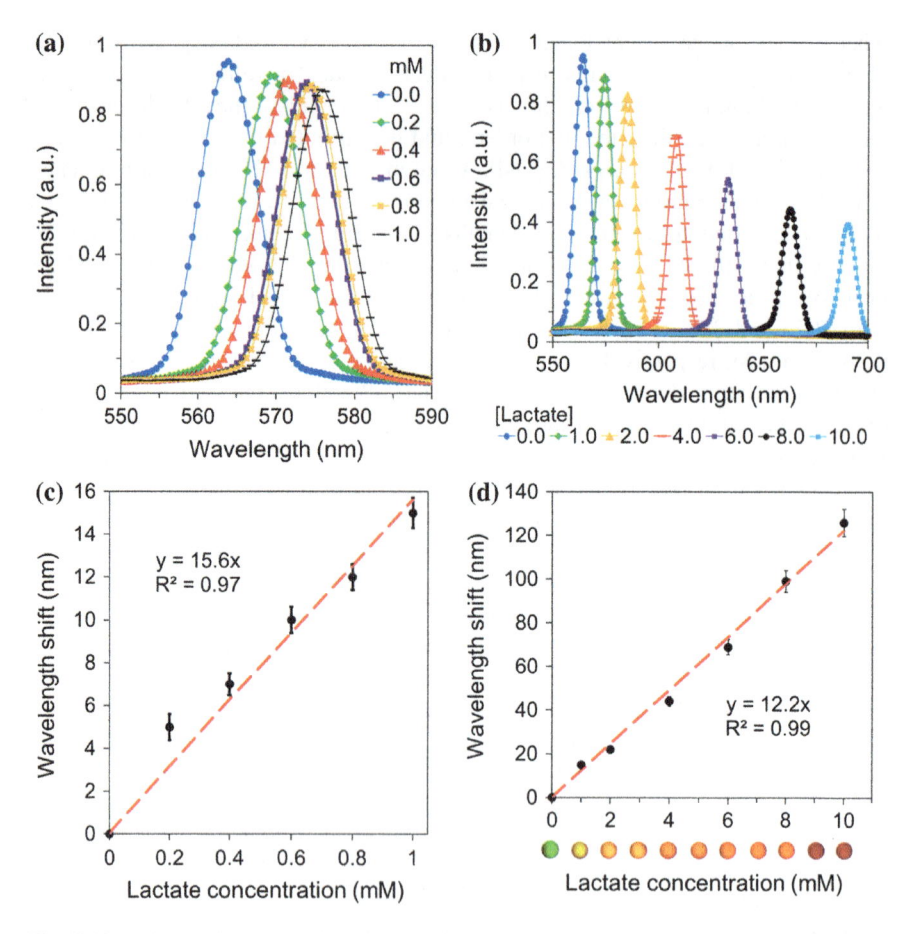

Fig. 5.11 Holographic sensor (20 mol% 3-APB) response to variation in the concentration of lactate at 24 °C in artificial urine. **a, c** An increase in the concentration (10.0 mM) of lactate in artificial urine (pH 7.40) shifted the Bragg peak from ∼563 to ∼689 nm, **b, d** Lower lactate concentrations (0.0–10.0 mM) shifted the peak from ∼563 to ∼578 nm (see *upper inset*) (n = 39). The *lower inset* shows the colorimetric readouts. Standard *error bars* represent three independent samples. Reprinted with permission from [29] Copyright 2014 The American Chemical Society

with boronic acid groups in the hydrogel. While at low concentrations (<1.0 mM) of lactate, a Bragg peak shift of ∼15 nm was measured (Fig. 5.11a), at high concentrations (10.0 mM), a shift of ∼125 nm was measured (Fig. 5.11b). Figure 5.11c, d shows the readouts over three trials, and the lower inset in Fig. 5.11d illustrates the colorimetric response. At lactate concentrations of ∼1.0 mM in urine [7], the corresponding interference of lactate shifts the Bragg peak by 15 nm. The measurement errors due to lactate interference for the diagnosis of glucosuria (10.0 mM) and urinary tract infections (<1.0 mM) were ∼4.2 and ∼13.6 %, respectively.

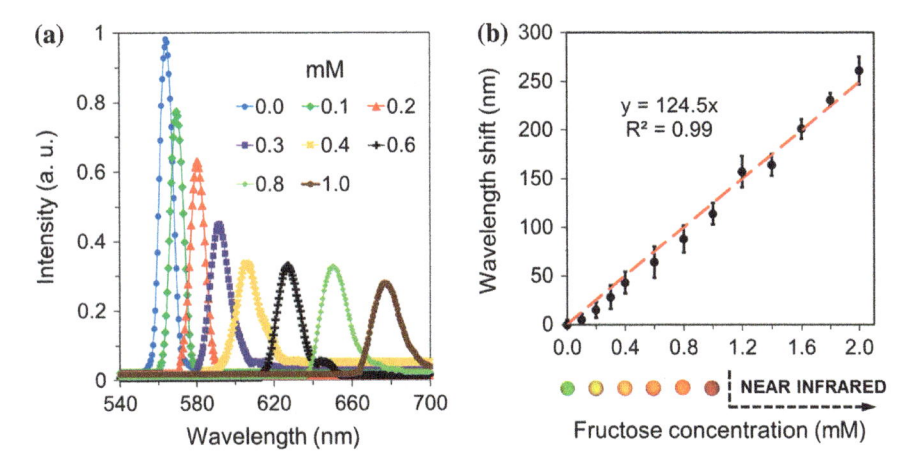

Fig. 5.12 Holographic sensor (20 mol% 3-APB) response to variation in the concentration of fructose at 24 °C in artificial urine. **a** An increase in the concentration (1.0 mM) of fructose in artificial urine shifted the Bragg peak from ∼565 to ∼678 nm (n = 39). **b** Fructose response up to 2.0 mM over three trials. The *inset* shows the colorimetric readouts. Standard *error bars* represent three independent samples. Reprinted with permission from [29] Copyright 2014 The American Chemical Society

Competitive 3-APB binding to fructose was evaluated. Normal fructose excretion in urine is 37.7 ± 23.0 μmol/day [51]. The concentration of fructose in urine may be elevated due to excessive dietary carbohydrate intake. Both natural and added sugar contain significant amounts of fructose [52], and dietary sugar consumption is proportional to the urinary excreted fructose [53] since a fraction of ingested fructose escapes hepatic metabolism and passes into the systematic circulation, where it is excreted in the urine [54]. To assess its interference, the holographic sensor (20 mol% 3-APB) was tested with different concentrations of fructose in artificial urine. Bragg peak shifts of ∼115 and ∼260 nm in the presence of 1.0 and 2.0 mM fructose were measured (Fig. 5.12a). The sensor response over three trials was also evaluated (Fig. 5.12b). Inset in Fig. 5.12b shows the colorimetric response.

5.5.4 Interference Due to Osmolality

By decreasing the IS of artificial urine samples (1.0 mM glucose, pH 7.4) through serial dilution, the effect of the salt concentration on the sensor response was assessed. As the IS of the artificial urine samples decreased from 250 to 25 mM, the Bragg peak red shifted by ∼75 nm due to an increase in Donnan osmotic pressure (Fig. 5.13).

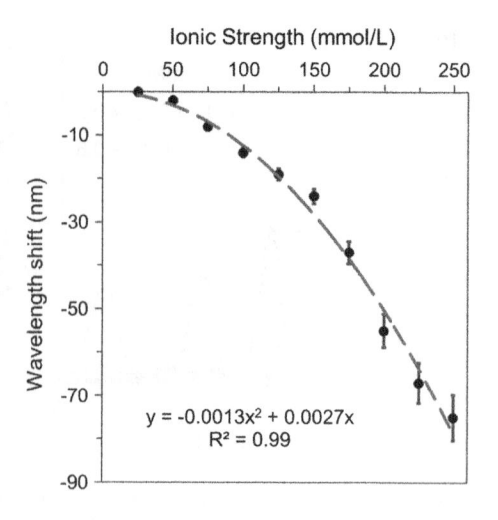

Fig. 5.13 The blue Bragg peak shifts of the holographic glucose sensor due to variation in osmolality. Standard *error bars* represent three independent samples (n = 30)

5.5.5 Tuning of the Wavelength Shift Range of the Holographic Glucose Sensor

The crosslinking density and the concentration of 3-APB of the holographic sensor can be varied to control the range of the Bragg peak shift. Figure 5.14a illustrates the Bragg peak shifts of the sensor that consisted of 5.0 mol% MBAAm and 20 mol % 3-APB as the concentrations of glucose were increased up to 50 mM in artificial urine. However, Fig. 5.14b shows the Bragg peak shifts of the sensor with reduced 3-APB concentration (10 mol%). Both of these approaches allowed sensing glucose up to and over 50 mM. In the approach involving the reduced concentration of 3-APB, the *cis* diols in tetrahedral coordination was the limiting factor, which also defined the sensitivity of the sensor.

5.5.6 Exposure Bath to Tune the Base Position of the Bragg Peak

Before exposing the hologram to laser light, any agent, which swells the hydrogel matrix will shrink the finished lattice spacing. This tuning mechanism is the opposite effect of that measured in the readout with the same agent. Glucose variation in the exposure bath during laser writing allowed adjusting the Ag^0 NP lattice spacing and tuned the base position of the Bragg peak. As the glucose concentration of the bath solution was increased from 0.5 to 4.0 mM, the Ag^0 NP distribution densities ($\sim \lambda/2$) were recorded at an increasing degree of swollen state of the polymer. After the laser exposure, the initial wavelength was measured in artificial urine solutions (pH 7.4) containing no glucose. The Bragg peak

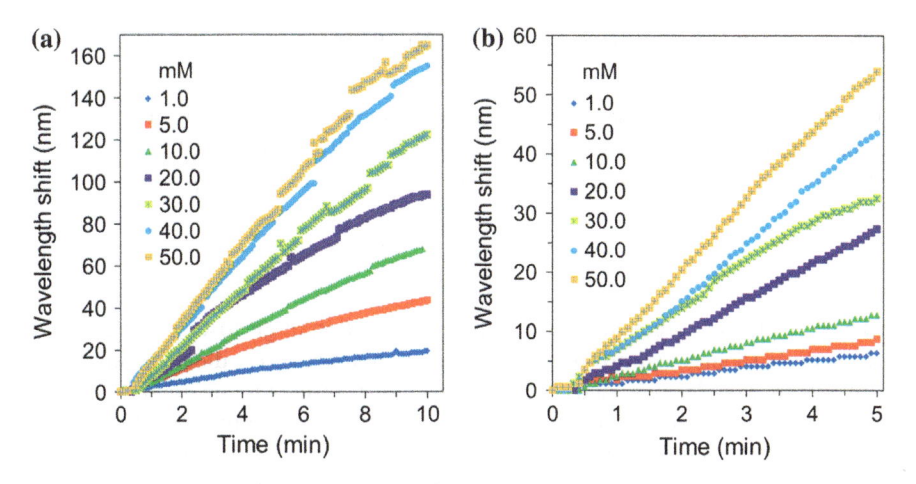

Fig. 5.14 Tuning the sensor's Bragg peak shift range by changing the concentration of the crosslinker and the 3-APB. The Bragg peak shifts of the holographic sensor due to variation in the concentration (1.0–50.0 mM) of glucose at pH 7.40 (n = 7) at 24 °C. The compositions of the sensors consisted of **a** AAm (75.0 mol%), MBAAm (5.0 mol%) and 3-APB (20.0 mol%), **b** AAm (85.0 mol%), MBAAm (5.0 mol%) and 3-APB (10.0 mol%)

($\sim$560 nm) shifted to shorter wavelengths ($\sim$430 nm) as the concentration of the glucose in the exposure bath was increased (Fig. 5.15). Hence, Ag^0 NP distribution densities with a lattice constant $<\lambda/2$ can be obtained, and the Bragg peak shift range of the sensor can be tuned.

Fig. 5.15 Tuning the sensor's base position of the Bragg peak by changing the composition of the exposure bath. Standard *error bars* represent three independent samples (n = 27)

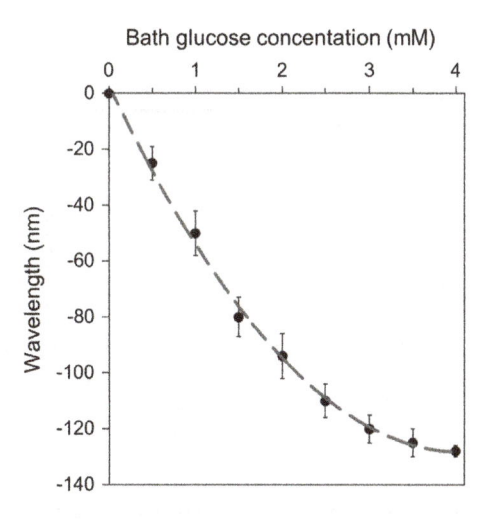

5.6 Kinetic Theory for Hydrogel Swelling

The capability for the glucose binding to occur is associated with the amount of *cis* diols available within the polymer matrix and the concentration of glucose in the sample measured. A change in the number of molecules bound $\eta(t)$ is:

$$\frac{d}{dt}\eta(t) \propto \beta(N - \eta(t)) = c_1\beta(N - \eta(t)) \tag{5.2}$$

where t is the measurement time, β is the concentration of glucose, N is the number of *cis* diols (bound and not bound), $\eta(t)$ is the number of *cis* diols already bound with glucose molecules, and c_1 is a constant. This expression implies that the concentration β is a constant, in other words, the amount of glucose molecules is in excess than the amount of *cis* diols and therefore, it is assumed that the concentration does not decrease. This relationship can be expressed as:

$$\eta(t) = N - c_2 e^{-c_1\beta t} \tag{5.3}$$

Figure 5.16a shows a typical simulated readout from a holographic sensor. When the Bragg peak measurement is taken from the poly(AAm-co-3-APB) matrix, the binding process may have already started, in such a case, the initial measurement is not $t = 0$ but an arbitrary t_i. The difference of $\eta(t)$ from a fixed t_i to the next measurement time t can be expressed as:

$$\Delta\eta(t) = \eta(t) - \eta(t_i) = c_2 e^{-c_1\beta t_i}\left(1 - e^{-c_1\beta(t-t_i)}\right) = c_3\left(1 - e^{-c_1\beta(t-t_i)}\right) \tag{5.4}$$

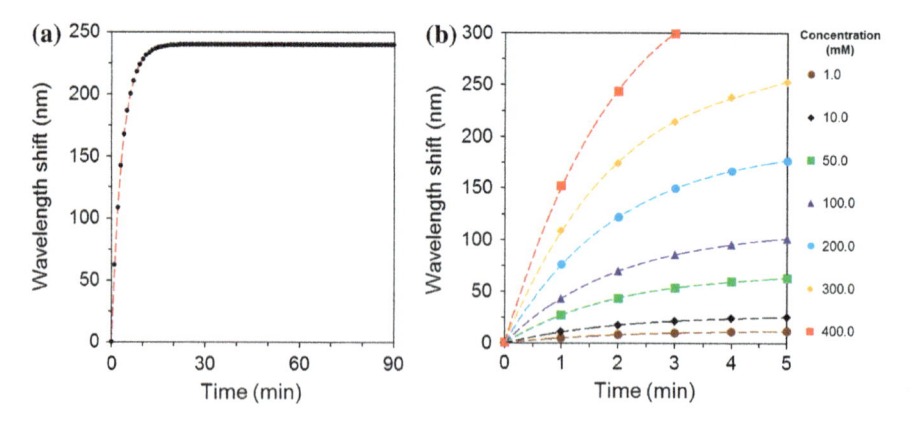

Fig. 5.16 The systematic approach to extrapolate the readouts using kinetics of hydrogel swelling. **a** A typical sensor response showing the correlation between time (min) and Bragg peak shift (nm). As the time increases, the Bragg peak shift saturates after a time point. **b** The initial calibration curves set for the measurements

The *cis* diol-glucose binding expands the volume of the hydrogel matrix normal to its underlying substrate. The matrix is constrained in its movement since it is attached to a plastic or glass substrate. When the system expands homogeneously normal to its substrate, a change in $\eta(t)$ alters the thickness x of the matrix:

$$\Delta\eta(t) \propto \Delta x = c_4 \Delta x \tag{5.5}$$

Equation (5.5) is based on the assumption that the *cis* diols are homogeneously distributed throughout the matrix. A change in the periodicity of the multilayer structure λ_h tilted by an angle θ with respect to the normal surface is:

$$\Delta\lambda_h \propto \Delta x \cos(\theta) = c_5 \Delta x \cos(\theta) \tag{5.6}$$

This expression implies that the entire matrix expands homogeneously, and the shift in the Bragg peak λ is proportional to λ_h:

$$\Delta\lambda \propto \Delta\lambda_h \propto \Delta x \cos(\theta) \propto \Delta\eta(t)\cos(\theta) \propto e^{-c_1\beta t_i}(1 - e^{-c_1\beta(t-t_i)})\cos(\theta) \tag{5.7}$$

Hence, for measurements developing on time with a fixed sample:

$$\lambda = \lambda_i + c_6\left(1 - e^{-c_1\beta(t-t_i)}\right) \tag{5.8}$$

If this curve has a Bragg peak position at the infinite λ_∞; $c_6 = \lambda_\infty - \lambda_i$, which can be expressed as:

$$\lambda = \lambda_\infty(1 - c_7 e^{-c_1\beta t_i}) \tag{5.9}$$

$$c_7 = \left(1 - \frac{\lambda_i}{\lambda_\infty}\right)e^{c_1\beta t_i} \tag{5.10}$$

Using Eq. (5.10), it is possible to fit a series of measurements developing in time and retaining specific features for all the constants except for c_7. The value of c_7 does not have a specific meaning since it involves t_i and λ_i, and although they are known values in the measurements, they have their associated uncertainties. When these values are fixed, for instance, with the initial (base) Bragg peak measurement and an initial time of $t = 0$, it implies no uncertainty in this specific measurement, and c_7 becomes a fixed value. When $t_i = 0$, the value $c_7 = 1 - \lambda_i/\lambda_\infty$. It can be inferred from this hypothesis that λ_∞ does not provide useful information since it is only related to N, but not to the concentration of glucose. Therefore, the final Bragg peak measurement saturates after an infinite time. When the time for multiple glucose measurements is constrained, the Bragg peak can correspond to different values at a specific time point while they all approach to an equilibrated Bragg peak λ_∞. Figure 5.16b illustrates the theoretical calibration curves for the initial and final measurement times. However, if the reading is taken too early or too late, it may not be possible to have a clear distinction between the measurements. By focusing on the constant c_1, the glucose concentration β can be correlated with a Bragg peak shift. The constant c_1 is generally related to the complexion capacity of the *cis* diol groups of 3-APB with glucose molecules.

When a Bragg peak shift in time (0–5 min) is measured, the rate of Bragg peak shift decreases as the time increases based on the decrease of *cis* diol groups in the tetrahedral coordination form. As the time lapses, the binding mechanism transforms from tetrahedral from to trigonal planar form, which is kinetically slower. Therefore, the rate of change of bound molecules $n(t)$ is proportional to the total number of glucose N_g molecules and the total number of *cis* diol groups N_f:

$$\frac{dn(t)}{dt} \propto \left(N_g - n(t)\right)\left(N_f - n(t)\right) = a\left(N_g - n(t)\right)\left(N_f - n(t)\right) \tag{5.11}$$

When the Bragg peak shift is proportional to the amount of binding of *cis* diol groups to glucose molecules, this expression can be reduced to:

$$\frac{d\lambda(t)}{dt} \propto \left(C_g - \Delta\lambda(t)\right)\left(C_f - \Delta\lambda(t)\right) = a\left(C_g - \Delta\lambda(t)\right)\left(C_f - \Delta\lambda(t)\right) \tag{5.12}$$

where C_g and C_f are constants with proportionality to *cis* diol and glucose groups present. The solution of this equation is nontrivial to be fitted with the experimental data; instead, the numerical solution of $d\lambda(t)/dt$ has been obtained from the data. Thus, two values were plotted, and a quadratic solution was fitted to the equation to obtain the variables C_g and C_f. As the solution was commutable (changing the order of the operands does not change the result), the minimum value was considered as C_g. Other factors such as a potential decrease in the elasticity of the polymer matrix might influence the projected decrease of the Bragg peak shift. Figure 5.17 shows three independent measurements with urine samples of diabetic patients at different glucose concentrations. The increase in the slope of the fit can be correlated with the concentration of carbohydrate (mainly glucose) in the urine samples.

5.7 Quantification of Glucose Concentration in Urine

Anonymised urine samples (n = 33) were collected from diabetic patients attending the Wolfson Diabetes and Endocrine Clinic (Addenbrooke's Hospital, Cambridge, UK) in February 2014 under the Human Tissue Act 2004 (c 30) of the UK. Urine samples were frozen immediately for glucose, fructose and lactate testing. Samples were thawed and kept at 4 °C, and centrifuged (1 min, 10,000 rpm) to remove any precipitates prior to testing. For holographic glucose sensor measurements, the urine samples had a pH of 5.96 ± 0.11 and it was adjusted to 7.40 by adding NaOH (2.0 M) while monitoring the pH using an electrochemical pH meter. The measurements of Bragg peak shifts from the holographic sensor allowed inferring the

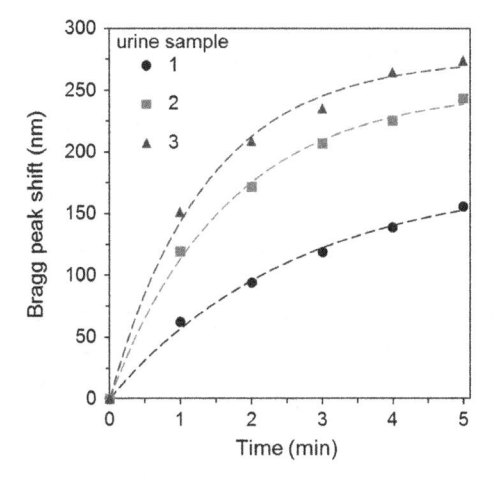

Fig. 5.17 Bragg peak measurements of the holographic sensor for three independent (random) urine samples of diabetic patients (pH 7.4). The increase in the slope shows an increase in the concentration of the carbohydrate in the urine samples. The sensor was reset in ∼10 s, and the Bragg peak baseline was 519 ± 5.8 nm

concentrations of glucose in urine samples. The values were correlated to the concentration of glucose found with Multistix® 10 SG (GOx method, Siemens) strips with CLINITEK Status® + Analyzer and the fully-automated Dimension® Clinical Chemistry System (hexokinase-glucose-6-phosphate dehydrogenase method (HK/G6P-DH), Siemens). Analyses with the urine strip tests were performed by briefly (∼1 s) dipping the Multistix® 10 SG test strip into the urine sample and ensuring that all test areas were moistened. While withdrawing the test strip from the urine sample, the edge of the test strip was wiped against the rim of the recipient to remove excess urine, followed by 1 s blot drying. After 60 s of reaction time, the test strip was analysed with CLINITEK Status® + Analyzer. As a control, the reaction colours of the test strips were also compared to the reference chart provided by the supplier. The sensitivity of HK/G6P-DH glucose assay was 0.056 mmol/L [55]. Hexokinase catalyses the phosphorylation of glucose in the presence of adenosine-5′-triphosphate (ATP) and Mg^{2+} ions to form glucose-6-phosphate (G-6-P) and adenosinediphosphate (ADP). G-6-P is then oxidised by glucose-6-phosphate dehydrogenase (G-6-PDH) in the presence of nicotinamide adenine dinucleotide (NAD^+) to produce 6-phosphogluconate and nicotinamide adenine dinucleotide hydride (NADH). One mole of NAD^+ is reduced to one mole of NADH for each mole of glucose present. The absorbance due to NADH (hence glucose concentration) was determined using a bichromatic (340 and 383 nm) endpoint technique. Reagents in HK/G6P-DH method included HK (15 U/mL, liquid, yeast sourced, well 1), G-6-PDH (30 U/mL, yeast sourced, wells 2), NAD^+ (8 mmol/L, well 3), ATP (15 mmol/L, well 4), Mg^{2+} ions (7.4 mmol/L, well 5), and stabiliser and buffer (well 6).

$$\text{Glucose} + \text{ATP} \xrightarrow{\text{HK}+\text{Mg}^{2+}} \text{Glucose-6-phosphate} + \text{ADP}$$

$$\text{Glucose-6-phosphate} + \text{NAD}^+ \xrightarrow{\text{G-6-PDH}} \text{6-phosphogluconate} + \text{NADH} + \text{H}^+$$

Assay Quality Control was achieved by analysing at least two levels of a serum-based material (Chem 1 Calibrator) with known glucose concentrations. The concentrations and their corresponding standard deviations were 3.4 ± 0.3 and 14.6 ± 0.7 mmol/L, respectively. The assay range was 0.1–27.8 mmol/L. The calibration scheme was based on three levels in triplicate. The assigned coefficients were $C_0 = 0.000$ and $C_1 = 0.880$. The Dimension$^{®}$ Clinical Chemistry system ran a sample (3 μL) along with reagent 1 (56 μL) and diluent (321 μL) at 37 °C. Sampling, reagent delivery, mixing and processing were performed by the Dimension$^{®}$ system. The samples with concentrations greater than 28 mmol/L were diluted (1:10, v/v) in PBS and retested. A limitation of the testing mechanism is that metronidazole (an antibiotic) of 2.34 μmol/L increases glucose readouts by 0.37 mmol/L at a glucose concentration of 1.11 mmol/L [56].

The inferred glucose concentrations showed an agreement between the holographic glucose sensor and Dimension$^{®}$ Clinical Chemistry System, while no false negatives were observed. The holographic glucose sensor had an improved correlation coefficient (R^2) of 0.79 as compared to Multistix$^{®}$ 10 SG (0.28) (Fig. 5.18). Multistix$^{®}$ 10 SG reported lower concentrations of glucose than Dimension$^{®}$

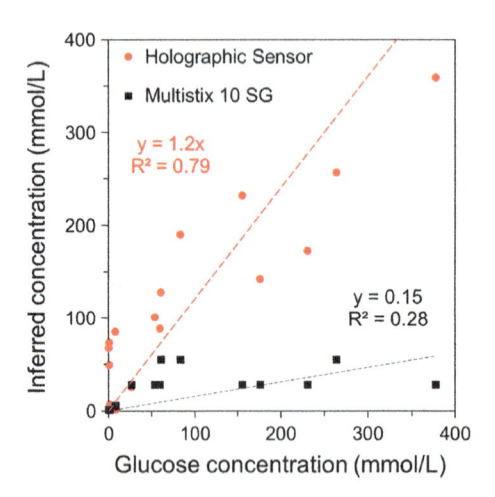

Fig. 5.18 Readouts of glucose concentrations using the holographic sensor in urinalysis as compared to Multistix$^{®}$ 10 SG read by CLINITEK Status$^{®}$, and Dimension$^{®}$ Clinical Chemistry System. The x-axis is based on the Dimension$^{®}$ system; and *black square* represents readouts from Multistix$^{®}$ 10 SG, and *red circle* represents inferred readouts from the holographic sensor. Reprinted with permission from [29] Copyright 2014 The American Chemical Society

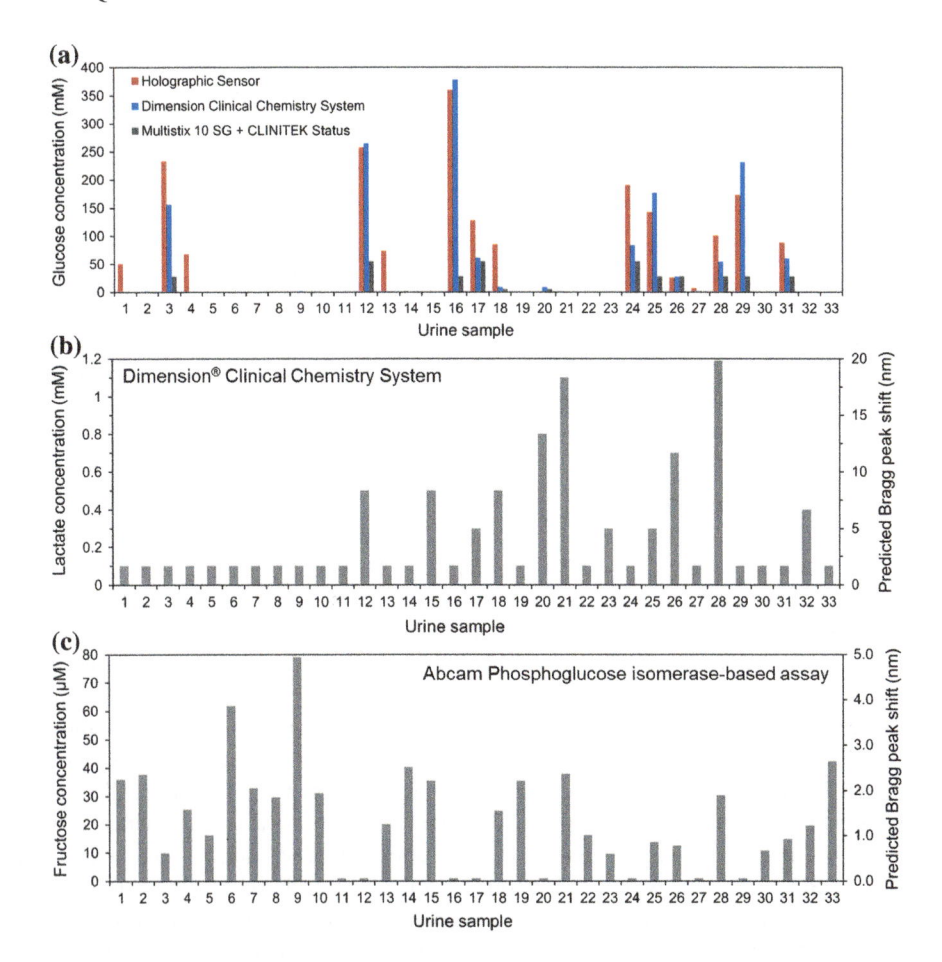

Fig. 5.19 Readouts for glucose, lactate and fructose concentrations in urinalysis. Measurements of **a** glucose, **b** lactate and **c** fructose concentrations in urine samples. The secondary axis shows the predicted Bragg peak shifts from the holographic sensor measurements in artificial urine. Data points represent individual urine specimens (n = 33)

Clinical Chemistry System and the holographic sensor. Figure 5.19a shows the data points obtained from the holographic sensor, Multistix® 10 SG strips with CLIN-ITEK Status® + Analyzer and Dimension® Clinical Chemistry System.

5.8 Lactate and Fructose Interference

The lactate assay is based on a modified Marbach and Weil method [57], which employs the oxidation of lactate to pyruvate, and its analytical sensitivity is <0.3 mmol/L. Rabbit muscle lactate dehydrogenase (LDH) catalyses the oxidation of L-lactate to pyruvate with simultaneous reduction of nicotinamide adenine dinucleotide (NAD^+). One mole of NAD^+ is converted to one mole of NADH for each mole (equivalent) of lactate present.

$$\text{L-lactate} + NAD^+ \xrightarrow{LDH} \text{Pyruvate} + NADH + H^+$$

Hydrazine is used to trap the pyruvate (hydrazone) as it is formed, thus driving the reaction to completion. The absorbance due to NADH is directly proportional to the concentration of lactate and is measured using a two-filter (340–383 nm) end point technique. Reagents included NAD^+ (2.0 µM, tablet, wells 1, 2), dihydrazine sulphate (3.7 µM, liquid, wells 3, 4), LDH (40 U, liquid, sourced from rabbit, wells 5, 6) and tris buffer (0.24 mM, liquid, wells 7, 8). Hydrating, diluting and mixing were automatically performed by the Dimension® System. Volumes of sample, reagent 1, 2, 3, 4 and diluent volume were 4, 158, 20, 75, 20, and 197 µL, respectively. Experiments were carried out at 37 °C. The assay range was 0.3–15.0 mmol/L and calibration scheme was based on 3 levels (n = 3). Calibrator concentrations were 0, 8 and 15 mmol/L. The assigned coefficients were $C_0 = -1.156$ and $C_1 = 0.0451$. Two levels of a quality control material with known lactic acid concentration were analysed without dilution.

The fructose assay was based on the enzymatic conversion of free fructose to β-glucose, which was then specifically converted to a product that reacted with 10-acetyl-3,7-dihydroxyphenoxazine (H_2O_2 probe, OxiRed™) to generate a colour (λ = 570 nm) [58]. The kit was stored at −20 °C and protected from light. The assay buffer was allowed to warm up to room temperature (24 °C) before use. All small vials were briefly centrifuged prior to opening. Phosphoglucose isomerase (EC 5.3.1.9, fructose converting enzyme) was only stable in ammonium sulphate (($NH_4)_2SO_4$) solution. A required amount (450 µl) for each assay (10 µL for each well) was centrifuged for 5 min at 10,000 rpm. The supernatant was removed and reconstituted with same volume of assay buffer. The enzyme mix was dissolved in 220 µL assay buffer separately. The samples (50 µL) were added into a 96-well plate. Fructose standard solution (100 mM) was diluted to 1 mM by adding 10 µL of fructose standard to 990 µL of assay buffer and mixed. The standard solution (1 mM) was serially diluted 1:2 (v/v) in assay buffer to give standard solutions (500, 250, 125 µmol/L), which were added in a series of wells. Enough reagent was mixed for the number of assays to be performed. For each well, reaction mix (50 µL) contained assay buffer (36 µL), OxiRed™ Probe (2 µL), enzyme mix (2 µL) and fructose converting enzyme (10 µL). The solution was mixed by a vortex mixer. Reaction mix (50 µL) was added to each well containing fructose standard and test samples. The resulting solution was mixed on the orbital plate shaker. The reaction was incubated for 30 min at 37 °C and protected from light using Al foil. Optical density $(OD)_{550\ nm}$ was measured for the colorimetric assay in a microplate reader. Due to the OxiRed™ probe, glucose generates background, which was subtracted by conducting a control without phosphoglucose isomerase in the reaction. The background was corrected by subtracting the value from the zero-

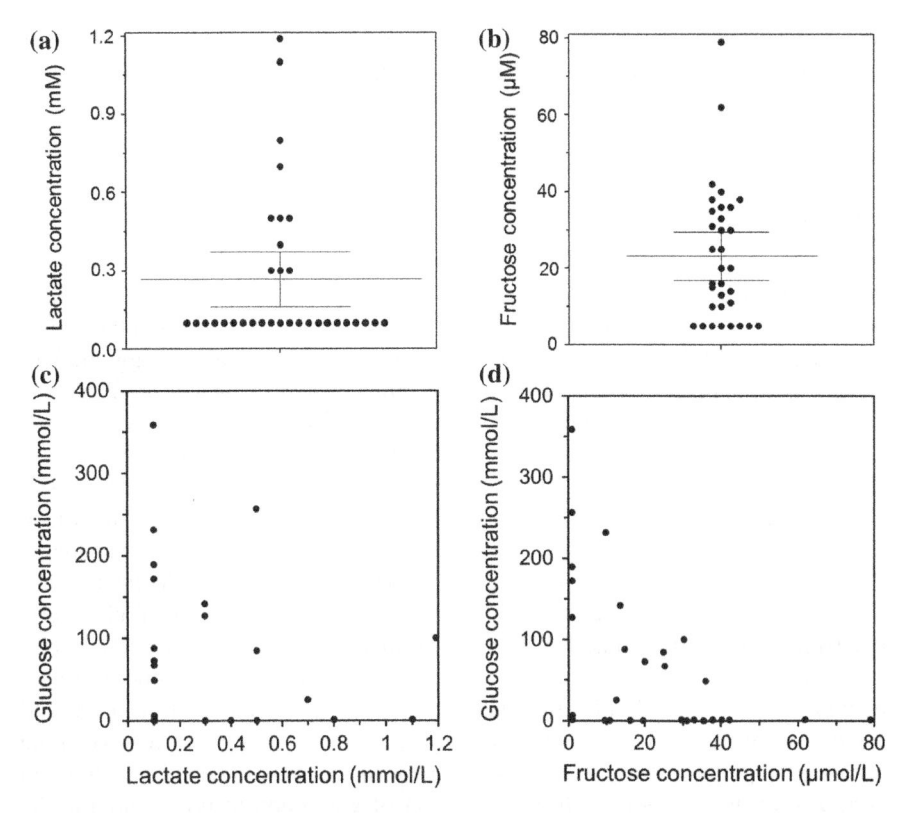

Fig. 5.20 Assessment of lactate and fructose interference in urinalysis. Measurements of **a** lactate **b** fructose concentrations *middle horizontal lines* in (**a**, **b**) are means. *Error bars* represent upper and lower 95 % confidence intervals of the mean. Comparison of **c** lactate and **d** fructose concentrations with inferred glucose concentrations from the holographic sensors. Reprinted with permission from [29] Copyright 2014 The American Chemical Society

fructose control from sample readouts. The fructose standard curve was plotted, and the fructose concentrations of test samples were read from the standard curve.

Figure 5.19b, c shows the lactate and fructose measurements for each sample. The secondary axes in Fig. 5.19b, c show the predicted Bragg peak shifts for lactate and fructose measurements, respectively. Figure 5.20a illustrates scatter dot plot of the measured lactate concentration values (μ = 0.27 mM, 95 % $CI_{low-high}$ = 0.16–0.37 mM), while Fig. 5.20b shows the measured fructose concentration values (μ = 23.18 μM, 95 % $CI_{low-high}$ = 16.93–29.43 μM). Such concentrations caused slight interferences of 3.48 ± 0.01 % (lactate) and 2.90 ± 0.44 % (fructose) in the presence of urine glucose. However, for the present analysis, it was redundant to calibrate the sensor due to lactate and fructose measurements since the interference due to these analytes was negligible. Figure 5.20c, d shows the inferred glucose values from the holographic sensor readouts plotted against lactate and fructose measurements using Dimension® Clinical Chemistry System and phosphoglucose isomerase-based assay (Abcam), respectively. The results show no correlation between the predicted glucose concentrations versus lactate and fructose assays.

5.9 Conclusions

A holographic sensor with point-of-care clinical applicability for diabetes screening or diagnosis of urinary tract infection was demonstrated. The holographic sensor was fabricated by using a single 6 ns laser pulse to produce Bragg gratings consisting of periodic Ag° NP distribution regions that are separated $\sim \lambda/2$ apart in a poly(AAm-co-3-APB) matrix. The sensor displayed Bragg peak shifts in the visible spectrum and near infrared, while also being suitable for multiple analyses. The shifts in the Bragg peak allowed predicted measurements over the physiological glucose concentrations up to 375.0 mM. When the *cis* diols of the sensor bound to the glucose molecules, the Bragg peak of the sensor shifted to longer wavelengths, which produced visual colour changes. Additionally, the interference due to lactate and fructose was evaluated. For the quantification of glucosuria (>10.0 mM glucose), the interference from normal urinary lactate and fructose were 1.57 and 0.32 % of the readouts, respectively. However, for monitoring of urinary tract infections (<1.0 mM glucose), the interference from normal urinary lactate and fructose were 6.21 and 1.33 %, respectively. Table 5.4 shows the selectivity of the holographic glucose sensor in urinalysis. The readouts were obtained within 5 min with a reset time of ~ 10 s. In the detection of glucosuria in urine samples, the holographic sensor showed improved performance as compared to Multistix® 10 SG strips with CLINITEK Status® + Analyzer, while showing comparable performance with the fully-automated Dimension® Clinical Chemistry System. In the realisation of the holographic glucose sensors, the pH of urine can be corrected using analytical methods, while the osmolality may be standardised using glucose/creatinine ratio. Furthermore, the *cis* diols of 3-APB in the holographic sensor bind

Table 5.4 Selectivity assessment of the holographic sensor in urine samples

	Concentration	λ_{shift} (nm)	Interference (%) <1.0 mM glucose (~ 85 nm)	Interference (%) 10.0 mM glucose (~ 355 nm)
Lactate	*Normal* 0.27 mM	5.7	6.21	1.57
	Abnormal 0.37 mM	6.7	7.22	1.84
Fructose	*Normal* 23.18 µM	1.16	1.33	0.32
	Abnormal 29.43 µM	1.47	1.68	0.41

to glucose, lactate and fructose molecules as well as other carbohydrates such as xylose, galactose, maltose, ribose, mannose, lactose and sucrose [18]. Urine consists of many carbohydrates including maltose, lactose, D-mannose, D-glucose, D-ribose, D-xylose, L-arabinose and D-galactose [59]. The Bragg peak shift in holographic sensor readouts can be expressed as:

$$\Delta\lambda_{tot} = \Delta\lambda_{glu} + \Delta\lambda_{lact} + \Delta\lambda_{fru} + \Delta\lambda_{carbo} - \Delta\lambda_{IS} \qquad (5.13)$$

where $\Delta\lambda_{lac}$, $\Delta\lambda_{fruc}$ and $\Delta\lambda_{carbo}$ are the red Bragg peak shifts due to glucose, lactate, and fructose and other carbohydrates, and $\Delta\lambda_{IS}$ is the blue Bragg peak shift due to increase in IS from a standard molarity reference. Since the Bragg peak shift readouts were a combination of many other carbohydrates, elimination of these interferents may be required for accurate analyses.

Glucose sensors comprising of optical transducers embedded into analyte-responsive materials are attractive for the development of healthcare monitoring systems [46, 60]. The advantages of optical sensors over traditional dyes [61], fluorescent molecules [62, 63] and electrochemical [64–67] assays are that they: (i) are not affected by electromagnetic fields, (ii) are label-free, (iii) enable sterile remote sensing, (iv) are amenable to miniaturisation and multiplexing, and (v) are able to be used in real-time continuous monitoring [68]. Notable optical sensors have included photonic structures such as plasmonic nanomaterials [69], hybrid nanogels [70], photonic crystalline colloidal arrays [71], and inverse opal hydrogels [72]. These analyte-responsive polymers have the added advantage of fine tuning through a change in periodic structure, index of refraction and/or localised surface plasmon resonance. Although these polymeric optical sensors can be microfabricated, self-assembled or a combination of both; there is currently no rapid, low-cost and generic sensor fabrication technique capable of producing narrow-band, uniform, reversible colorimetric readouts with a high-tunability range. The chapter showed a holographic sensor, which was fabricated using laser light to allow forming Bragg gratings in a rapid manner. While the range of spectral readouts was finely controlled by changing the wavelength of the laser light and the chemistry of the exposure bath; the sensitivity of the photonic sensor was modulated by varying the concentration of the crosslinker and the functional groups in the hydrogel matrix. The kinetic theory developed in this chapter is based on the correlation of slope with the final readout, and it is the first proposed methodology for inferring concentrations of target analytes within minutes. This approach can be applied to any hydrogel-based sensor or polymeric nanoparticle-based drugs for the optimisation of binding kinetics. Many diagnostic approaches have utilised boronic acid derivatives in the development of fluorescence, colorimetric, electrochemical and optical sensors [46, 60, 73, 74]. The advantage of the present study over these sensors is that holographic sensors produced by the rapid photochemical patterning offers routes to incorporate two or 3D images into the hydrogel matrix. For example, photomasks or 3D objects can be used during laser writing to produce user-friendly fool-proof text/quantity-reporting capabilities. The holographic sensing platform has flexibility in controlling the angle of off-axis diffraction precisely as well as the diffraction pattern, while offering a narrow-band response for semi-quantitative visual colorimetric readouts, and fully-quantitative readouts. Nanosecond laser writing of holographic sensors has the potential for producing equipment-free, reusable and scalable analytical devices. Additionally, the pulsed laser writing can allow patterning other nanostructures [75, 76]. The glucose sensors demonstrated in this chapter can be multiplexed with pH and metal ion sensors [2, 77–79]. These sensors can be printed in array format and multiplexed using microfluidics [80–86], and readouts may be obtained by smartphone applications and wearable devices [87, 88]. The technology is applicable to in vitro diagnostics as well as implantable in vivo sensors since it can operate in both visible and infrared regions. It is envisioned that holographic glucose sensors will enable the development of colorimetric tests for screening diabetes and urinary tract infections at clinical settings and point-of-care.

References

1. Zhang CJ, Losego MD, Braun PV (2013) Hydrogel-based glucose sensors: effects of phenylboronic acid chemical structure on response. Chem Mater 25(15):3239–3250. doi:10. 1021/Cm401738p
2. Yetisen AK, Butt H, da Cruz Vasconcellos F, Montelongo Y, Davidson CAB, Blyth J, Chan L, Carmody JB, Vignolini S, Steiner U, Baumberg JJ, Wilkinson TD, Lowe CR (2014) Light-directed writing of chemically tunable narrow-band holographic sensors. Adv Opt Mater 2 (3):250–254. doi:10.1002/adom.201300375
3. Danaei G, Finucane MM, Lu Y, Singh GM, Cowan MJ, Paciorek CJ, Lin JK, Farzadfar F, Khang YH, Stevens GA, Rao M, Ali MK, Riley LM, Robinson CA, Ezzati M, Global Burden of Metabolic Risk Factors of Chronic Diseases Collaborating G (2011) National, regional, and global trends in fasting plasma glucose and diabetes prevalence since 1980: systematic analysis of health examination surveys and epidemiological studies with 370 country-years and 2.7 million participants. Lancet 378(9785):31–40. doi:10.1016/S0140-6736(11)60679-X
4. Atkinson MA, Eisenbarth GS, Michels AW (2014) Type 1 diabetes. Lancet 383(9911):69–82. doi:10.1016/S0140-6736(13)60591-7
5. Diabetes Atlas (2013) 6 edn. International Diabetes Federation
6. Ravallion M (2013) How long will it take to lift one billion people out of poverty? World Bank Res Obser 28(2):139–158. doi:10.1093/Wbro/Lkt003
7. Kratz A, Ferraro M, Sluss PM, Lewandrowski KB (2004) Case records of the Massachusetts General Hospital. Weekly clinicopathological exercises. Laboratory reference values. N Engl J Med 351(15):1548–1563. doi:10.1056/NEJMcpc049016
8. Kabilan S, Marshall AJ, Sartain FK, Lee MC, Hussain A, Yang XP, Blyth J, Karangu N, James K, Zeng J, Smith D, Domschke A, Lowe CR (2005) Holographic glucose sensors. Biosens Bioelectron 20(8):1602–1610. doi:10.1016/j.bios.2004.07.005
9. Hisamitsu I, Kataoka K, Okano T, Sakurai Y (1997) Glucose-responsive gel from phenylborate polymer and poly (vinyl alcohol): prompt response at physiological pH through the interaction of borate with amino group in the gel. Pharm Res 14(3):289–293. doi:10.1023/A: 1012033718302
10. Springsteen G, Wang BH (2002) A detailed examination of boronic acid-diol complexation. Tetrahedron 58(26):5291–5300. doi:10.1016/S0040-4020(02)00489-1

11. Asher SA, Alexeev VL, Goponenko AV, Sharma AC, Lednev IK, Wilcox CS, Finegold DN (2003) Photonic crystal carbohydrate sensors: low ionic strength sugar sensing. J Am Chem Soc 125(11):3322–3329. doi:10.1021/ja021037h

12. Lowe CR, Davidson CAB, Blyth J, Kabilan S, Marshall AJ, Madrigal Gonzalez B, James AP (2003) Method of detecting an analyte in a fluid. WO Patent Application 2003087899

13. Lee MC, Kabilan S, Hussain A, Yang X, Blyth J, Lowe CR (2004) Glucose-sensitive holographic sensors for monitoring bacterial growth. Anal Chem 76(19):5748–5755. doi:10. 1021/ac049334n

14. Sartain FK, Yang X, Lowe CR (2006) Holographic lactate sensor. Anal Chem 78(16): 5664–5670. doi:10.1021/ac060416g

15. Kraiskii AV, Postnikov VA, Sultanov TT, Khamidul in AV (2010) Holographic sensors for diagnostics of solution components. IEEE J Quantum Electron 40(2):178–182. doi:10.1070/ Qe2010v040n02abeh014169

16. Lowe CR, Blyth J, Kabilan S, Hussain A, Yang XP, Sartain FK, Lee MC (2004) Holographic Sensor. WO Patent Application 2004081624 A1

17. Yang X, Lee MC, Sartain F, Pan X, Lowe CR (2006) Designed boronate ligands for glucose-selective holographic sensors. Chemistry 12(33):8491–8497. doi:10.1002/chem.200600442

18. Dean KES, Horgan AM, Marshall AJ, Kabilan S, Pritchard J (2006) Selective holographic detection of glucose using tertiary amines. Chem Commun 33:3507–3509. doi:10.1039/ B605778k

19. Kabilan S, Lee MC, Horgan AM, Medlock KES (2007) Novel boronate complex and its use in a glucose sensor. WO Patent Application 2007054689:A1

20. Worsley GJ, Tourniaire GA, Medlock KE, Sartain FK, Harmer HE, Thatcher M, Horgan AM, Pritchard J (2007) Continuous blood glucose monitoring with a thin-film optical sensor. Clin Chem 53(10):1820–1826. doi:10.1373/clinchem.2007.091629

21. Worsley GJ, Tourniaire GA, Medlock KE, Sartain FK, Harmer HE, Thatcher M, Horgan AM, Pritchard J (2008) Measurement of glucose in blood with a phenylboronic acid optical sensor. J Diabetes Sci Technol 2(2):213–220. doi:10.1177/193229680800200207

22. Farandos NM, Yetisen AK, Monteiro MJ, Lowe CR, Yun SH (2014) Contact lens sensors in ocular diagnostics. Adv Healthc Mater. doi:10.1002/adhm.201400504

23. Domschke A, Kabilan S, Anand R, Caines M, Fetter D, Griffith P, James K, Karangu N, Smith D, Vargas M, Zeng J, Hussain A, Yang XP, Blyth J, Mueller A, Herbrechtsmeier P, Lowe CR (2004) Holographic sensors in contact lenses for minimally-invasive glucose measurements. Proc IEEE Sens 3:1320–1323. doi:10.1109/Icsens.2004.1426425

24. Domschke A, March WF, Kabilan S, Lowe C (2006) Initial clinical testing of a holographic non-invasive contact lens glucose sensor. Diabetes Technol Ther 8(1):89–93. doi:10.1089/dia. 2006.8.89

25. Lowe CR, Kabilan S, Blyth J, Domschke A, Smith D, Karangu N (2005) Ophthalmic device comprising a holographic sensor. 2005031442 A1 (Application)

26. Burles B, Millington RB, Lowe CR, Kabilan S, Blyth J (2012) Ophthalmic device comprising a holographic sensor. US Patent 8,241,574 B2

27. Baca JT, Finegold DN, Asher SA (2007) Tear glucose analysis for the noninvasive detection and monitoring of diabetes mellitus. Ocul Surf 5(4):280–293. doi:10.1016/S1542-0124(12) 70094-0

28. Yang XP, Pan XH, Blyth J, Lowe CR (2008) Towards the real-time monitoring of glucose in tear fluid: holographic glucose sensors with reduced interference from lactate and pH. Biosens Bioelectron 23(6):899–905. doi:10.1016/j.bios.2007.09.016

29. Yetisen AK, Montelongo Y, da Cruz Vasconcellos F, Martinez-Hurtado JL, Neupane S, Butt H, Qasim MM, Blyth J, Burling K, Carmody JB, Evans M, Wilkinson TD, Kubota LT, Monteiro MJ, Lowe CR (2014) Reusable, robust, and accurate laser-generated photonic nanosensor. Nano Lett 14(6):3587–3593. doi:10.1021/nl5012504

30. Simerville JA, Maxted WC, Pahira JJ (2005) Urinalysis: a comprehensive review. Am Fam Physician 71(6):1153–1162

31. Scholl-Burgi S, Santer R, Ehrich JHH (2004) Long-term outcome of renal glucosuria type 0: the original patient and his natural history. Nephrol Dial Transpl 19(9):2394–2396. doi:10. 1093/Ndt/Gfh366
32. Ferrannini E (2011) Learning from glycosuria. Diabetes 60(3):695–696. doi:10.2337/Db10-1667
33. Scherstén B, Dahlqvist A, Fritz H, Köhler L, Westlund L (1968) Screening for bacteriuria with a test paper for glucose. JAMA 204(3):205–208. doi:10.1001/jama.1968.03140160015004
34. Wang J (2008) Electrochemical glucose biosensors. Chem Rev 108(2):814–825. doi:10.1021/cr068123a
35. Heller A, Feldman B (2008) Electrochemical glucose sensors and their applications in diabetes management. Chem Rev 108(7):2482–2505. doi:10.1021/Cr068069y
36. Davies MJ, Williams DR, Metcalfe J, Day JL (1993) Community screening for non-insulin-dependent diabetes mellitus: self-testing for post-prandial glycosuria. Q J Med 86 (10):677–684
37. Hanson RL, Nelson RG, McCance DR, Beart JA, Charles MA, Pettitt DJ, Knowler WC (1993) Comparison of screening tests for non-insulin-dependent diabetes mellitus. Arch Intern Med 153(18):2133–2140. doi:10.1001/archinte.1993.00410180083010
38. Friderichsen B, Maunsbach M (1997) Glycosuric tests should not be employed in population screenings for NIDDM. J Public Health 19(1):55–60
39. Engelgau MM, Narayan KM, Herman WH (2000) Screening for type 2 diabetes. Diabetes Care 23(10):1563–1580. doi:10.2337/diacare.23.10.1563
40. Feldman JM, Kelley WN, Lebovitz HE (1970) Inhibition of glucose oxidase paper tests by reducing metabolites. Diabetes 19(5):337–343. doi:10.2337/diab.19.5.337
41. Rotblatt MD, Koda-Kimble MA (1987) Review of drug interference with urine glucose tests. Diabetes Care 10(1):103–110. doi:10.2337/diacare.10.1.103
42. van der Sande MAB, Walraven GEL, Bailey R, Rowley JTF, Banya WAS, Nyan OA, Faal H, Ceesay SM, Milligan PJM, McAdam KPWJ (1999) Is there a role for glycosuria testing in sub-Saharan Africa? Trop Med Int Health 4(7):506–513. doi:10.1046/j.1365-3156.1999. 00430.x
43. Wei OY, Teece S (2006) Urine dipsticks in screening for diabetes mellitus. Emerg Med J 23 (2):139–140. doi:10.1136/emj.2005.033456
44. Chemstrip 10 MD package insert, cobas, Roche (2013)
45. Brooks T, Keevil CW (1997) A simple artificial urine for the growth of urinary pathogens. Lett Appl Microbiol 24(3):203–206. doi:10.1046/j.1472-765X.1997.00378.x
46. Guan Y, Zhang YJ (2013) Boronic acid-containing hydrogels: synthesis and their applications. Chem Soc Rev 42(20):8106–8121. doi:10.1039/C3cs60152h
47. Fritz H, Köhler L, Scherstén B (1969) Assessment of subnormal urinary glucose as an indicator of bacteriuria in population studies: an investigation of 3,911 subjects between the ages of four and sixty-five years. Acta Med Scand Berlin 504:1–39
48. Bensman A, Dunand O, Ulinski T (2009) Urinary tract infections. In: Avner E, Harmon W, Niaudet P, Yoshikawa N (eds) Pediatric nephrology, 6th edn. Springer, Berlin, pp 1297–1310. doi:10.1007/978-3-540-76341-3_54
49. Wang A, Nizran P, Malone MA, Riley T (2013) Urinary tract infections. Prim Care 40 (3):687–706. doi:10.1016/j.pop.2013.06.005
50. Talasniemi JP, Pennanen S, Savolainen H, Niskanen L, Llesivuori J (2008) Analytical investigation: assay of D-lactate in diabetic plasma and urine. Clin Biochem 41 (13):1099–1103. doi:10.1016/j.clinbiochem.2008.06.011
51. Kawasaki T, Akanuma H, Yamanouchi T (2002) Increased fructose concentrations in blood and urine in patients with diabetes. Diabetes Care 25(2):353–357. doi:10.2337/diacare.25.2. 353
52. Luceri C, Caderni G, Lodovici M, Spagnesi MT, Monserrat C, Lancioni L, Dolara P (1996) Urinary excretion of sucrose and fructose as a predictor of sucrose intake in dietary intervention studies. Cancer Epidem Biomar 5(3):167–171

53. Johner SA, Libuda L, Shi L, Retzlaff A, Joslowski G, Remer T (2010) Urinary fructose: a potential biomarker for dietary fructose intake in children. Eur J Clin Nutr 64(11):1365–1370. doi:10.1038/ejcn.2010.160

54. Tasevska N, Runswick SA, Welch AA, McTaggart A, Bingham SA (2009) Urinary sugars biomarker relates better to extrinsic than to intrinsic sugars intake in a metabolic study with volunteers consuming their normal diet. Eur J Clin Nutr 63(5):653–659. doi:10.1038/ejcn.2008.21

55. Richterich R, Dauwalder H (1971) Determination of plasma glucose by hexokinase-glucose-6-phosphate dehydrogenase method. Schweiz Med Wochenschr 101(17):615–618

56. Booklet, Glucose Assay, Dimension Clinical Chemistry System, Siemens (2014)

57. Marbach EP, Weil MH (1967) Rapid enzymatic measurement of blood lactate and pyruvate use and significance of metaphosphoric acid as a common precipitant. Clin Chem 13 (4):314–325

58. Protocol, Fructose Assay Kit (ab83380), Abcam (2014)

59. Jin LJ, Li SF (1999) Screening of carbohydrates in urine by capillary electrophoresis. Electrophoresis 20(17):3450–3454. doi:10.1002/(SICI)1522-2683(19991101)20:17<3450:AID-ELPS3450>3.0.CO;2-G

60. Wu Q, Wang L, Yu H, Wang J, Chen Z (2011) Organization of glucose-responsive systems and their properties. Chem Rev 111(12):7855–7875. doi:10.1021/cr200027j

61. Bankar SB, Bule MV, Singhal RS, Ananthanarayan L (2009) Glucose oxidase—an overview. Biotechnol Adv 27(4):489–501. doi:10.1016/j.biotechadv.2009.04.003

62. Liu Y, Deng C, Tang L, Qin A, Hu R, Sun JZ, Tang BZ (2011) Specific detection of D-glucose by a tetraphenylethene-based fluorescent sensor. J Am Chem Soc 133(4):660–663. doi:10.1021/ja107086y

63. Cummins BM, Garza JT, Cote GL (2013) Optimization of a Concanavalin A-based glucose sensor using fluorescence anisotropy. Anal Chem 85(11):5397–5404. doi:10.1021/ac303689j

64. Zhai D, Liu B, Shi Y, Pan L, Wang Y, Li W, Zhang R, Yu G (2013) Highly sensitive glucose sensor based on pt nanoparticle/polyaniline hydrogel heterostructures. ACS Nano 7(4):3540–3546. doi:10.1021/nn400482d

65. Invernale MA, Tang BC, York RL, Le L, Hou DY, Anderson DG (2014) Microneedle electrodes toward an amperometric glucose-sensing smart patch. Adv healthc mater 3 (3):338–342. doi:10.1002/adhm.201300142

66. Guo C, Huo H, Han X, Xu C, Li H (2014) Ni/CdS bifunctional Ti@TiO2 core-shell nanowire electrode for high-performance nonenzymatic glucose sensing. Anal Chem 86(1):876–883. doi:10.1021/ac4034467

67. Wooten M, Karra S, Zhang M, Gorski W (2014) On the direct electron transfer, sensing, and enzyme activity in the glucose oxidase/carbon nanotubes system. Anal Chem 86(1):752–757. doi:10.1021/ac403250w

68. Steiner MS, Duerkop A, Wolfbeis OS (2011) Optical methods for sensing glucose. Chem Soc Rev 40(9):4805–4839. doi:10.1039/c1cs15063d

69. He H, Xu X, Wu H, Jin Y (2012) Enzymatic plasmonic engineering of Ag/Au bimetallic nanoshells and their use for sensitive optical glucose sensing. Adv Mater 24(13):1736–1740. doi:10.1002/adma.201104678

70. Wu W, Mitra N, Yan EC, Zhou S (2010) Multifunctional hybrid nanogel for integration of optical glucose sensing and self-regulated insulin release at physiological pH. ACS Nano 4 (8):4831–4839. doi:10.1021/nn1008319

71. Muscatello MM, Stunja LE, Asher SA (2009) Polymerized crystalline colloidal array sensing of high glucose concentrations. Anal Chem 81(12):4978–4986. doi:10.1021/ac900006x

72. Honda M, Kataoka K, Seki T, Takeoka Y (2009) Confined stimuli-responsive polymer gel in inverse opal polymer membrane for colorimetric glucose sensor. Langmuir 25(14):8349–8356. doi:10.1021/la804262b

73. Lee YJ, Pruzinsky SA, Braun PV (2004) Glucose-sensitive inverse opal hydrogels: analysis of optical diffraction response. Langmuir 20(8):3096–3106. doi:10.1021/la035555x

74. Tierney S, Falch BMH, Hjelme DR, Stokke BT (2009) Determination of glucose levels using a functionalized hydrogel-optical fiber biosensor: toward continuous monitoring of blood glucose in vivo. Anal Chem 81(9):3630–3636. doi:10.1021/Ac900019k

75. Deng S, Yetisen AK, Jiang K, Butt H (2014) Computational modelling of a graphene Fresnel lens on different substrates. RSC Adv 4(57):30050–30058. doi:10.1039/C4ra03991b

76. Kong X-T, Butt H, Yetisen AK, Kangwanwatana C, Montelongo Y, Deng S, Cruz Vasconcellos Fd, Qasim MM, Wilkinson TD, Dai Q (2014) Enhanced reflection from inverse tapered nanocone arrays. Appl Phys Lett 105(5):053108. doi:10.1063/1.4892580

77. Tsangarides CP, Yetisen AK, da Cruz Vasconcellos F, Montelongo Y, Qasim MM, Wilkinson TD, Lowe CR, Butt H (2014) Computational modelling and characterisation of nanoparticle-based tuneable photonic crystal sensors. RSC Adv 4(21):10454–10461. doi:10.1039/C3RA47984F

78. Yetisen AK, Qasim MM, Nosheen S, Wilkinson TD, Lowe CR (2014) Pulsed laser writing of holographic nanosensors. J Mater Chem C 2(18):3569–3576. doi:10.1039/C3tc32507e

79. Yetisen AK, Montelongo Y, Qasim MM, Butt H, Wilkinson TD, Monteiro MJ, Lowe CR, Yun SH (2014) Nanocrystal Bragg grating sensor for colorimetric detection of metal ions (under review)

80. Yetisen AK, Naydenova I, Vasconcellos FC, Blyth J, Lowe CR (2014) Holographic sensors: three-dimensional analyte-sensitive nanostructures and their applications. Chem Rev 114 (20):10654–10696. doi:10.1021/cr500116a

81. Akram MS, Daly R, Vasconcellos FC, Yetisen AK, Hutchings I, Hall EAH (2015) Applications of paper-based diagnostics. In: Castillo-Leon J, Svendsen WE (eds) Lab-on-a-chip devices and micro-total analysis systems. Springer, New York

82. Volpatti LR, Yetisen AK (2014) Commercialization of microfluidic devices. Trends Biotechnol 32(7):347–350. doi:10.1016/j.tibtech.2014.04.010

83. Yetisen AK, Akram MS, Lowe CR (2013) Paper-based microfluidic point-of-care diagnostic devices. Lab Chip 13(12):2210–2251. doi:10.1039/c3lc50169h

84. Yetisen AK, Volpatti LR (2014) Patent protection and licensing in microfluidics. Lab Chip 14 (13):2217–2225. doi:10.1039/c4lc00399c

85. Vasconcellos FD, Yetisen AK, Montelongo Y, Butt H, Grigore A, Davidson CAB, Blyth J, Monteiro MJ, Wilkinson TD, Lowe CR (2014) Printable surface holograms via laser ablation. ACS Photonics 1(6):489–495. doi:10.1021/Ph400149m

86. Yetisen AK, Jiang L, Cooper JR, Qin Y, Palanivelu R, Zohar Y (2011) A microsystem-based assay for studying pollen tube guidance in plant reproduction. J Micromech Microeng 21 (5):054018. doi:10.1088/0960-1317/21/5/054018

87. Yetisen AK, Martinez-Hurtado JL, da Cruz Vasconcellos F, Simsekler MC, Akram MS, Lowe CR (2014) The regulation of mobile medical applications. Lab Chip 14(5):833–840. doi:10.1039/c3lc51235e

88. Yetisen AK, Martinez-Hurtado JL, Garcia-Melendrez A, Vasconcellos FC, Lowe CR (2014) A smartphone algorithm with inter-phone repeatability for the analysis of colorimetric tests. Sens Actuators B 196:156–160. doi:10.1016/j.snb.2014.01.077

Chapter 6
Mobile Medical Applications

The development of medical smartphone applications (apps) can allow quantification of rapid diagnostics at point-of-care and enable clinical data collection in real time. Mobile medical apps can reduce the erroneous subjective readouts, and create a standard readout platform with connectivity options at low cost. This chapter demonstrates the development of an app algorithm that utilises the camera of the Android and iPhone smartphones to read colorimetric tests. This smartphone app can be used with dipsticks, lateral-flow and flow-through assays as well as aqueous colorimetric tests that are typically read by spectrophotometers or microplate readers. The mobile app was designed to provide on-site quantitative screening when rapid diagnosis is needed. The utility of the smartphone app was demonstrated through quantifying pH, the concentrations of protein and glucose in commercial urine test strips, which had linear responses in the ranges of 5.0–9.0, 15–100 and 50–300 mg/dL, respectively. The app can be adapted for semi-quantitative analysis of commercial colorimetric tests, rendering it an inexpensive and accessible alternative to more costly commercial readers.

6.1 Global Health and Mobile Medical Applications

Decentralisation of healthcare through low-cost and highly portable point-of-care diagnostic devices has the potential to revolutionise current limitations in patient screening, particularly in the developing world, where the diagnosis is hindered by inadequate infrastructure and shortages in skilled healthcare workers [1, 2]. Overcoming such challenges by developing accessible diagnostics could reduce the large burden of disease [3, 4]. Currently, diagnostic devices such as strip tests (dipsticks) and lateral-flow tests are widely used to measure the concentration or detect the presence of various target analytes [5]. Such tests are employed for urinalysis, immunoassays, veterinary screening, food quality tests, environmental monitoring, biothreat detection and drug abuse screening. Dipstick and lateral-flow test formats are ubiquitous because of their portability, compactness, ease-of-use and equipment-free approach; rendering them universally applicable platforms for simple,

© Springer International Publishing Switzerland 2015
A.K. Yetisen, *Holographic Sensors*, Springer Theses,
DOI 10.1007/978-3-319-13584-7_6

multiplexed, qualitative or semi-quantitative low-cost point-of-care applications [6]. Whilst historically the lateral-flow format has been optimised for qualitative point-of-care diagnostics, the dipstick format has been designed for semi-quantitative measurements. For example, urine dipsticks are widely used in clinical practice in screening for renal, urinary, hepatic and metabolic disorders. Table 6.1 shows commercial semi-quantitative urine dipsticks in the market. Colorimetric tests are typically read by comparing the developed reaction zones with a reference chart. However, subjective interpretation may result in erroneous diagnosis, limiting the accuracy of colorimetric tests. To increase the accuracy of the measurement, colorimetric tests can be analysed using benchtop equipment such as spectrophotometers, or automated test-specific readers such as CLINITEK Status® + Analyzer (Siemens) or Urisys 1100® Urine Analyzer (Roche). Recently, these devices have been offered with connectivity options including data management solutions, such as data transfer through serial, Ethernet and wireless connection, barcode data entry, and support for healthcare connectivity protocols such as Health Level Seven (HL7) and point-of-care testing POCT1-A2 standards. However, these dipstick readers have a number of limitations that reduce their utility in resource-poor settings: (i) High retail price (>$1,000), (ii) requirement for reader-specific test strips, (iii) poor portability, and (iv) external power supply requirement.

The high mobile phone penetration and rapidly growing telecommunications infrastructure represent an unprecedented opportunity for reading and transferring point-of-care diagnostic data [7]. Global mobile-cellular subscriptions have grown 70 % over the last 5 years, reaching 7.3 billion as of 2014 (Fig. 6.1) [8]. Hence, taking advantage of the mobile phone infrastructure to monitor health conditions and the environment will provide low-cost screening for existing and emerging diseases, and improve the diagnostics at point-of-care setting. In telemedicine, a healthcare worker can capture the image of the rapid test (e.g. colorimetric) and send it to a server at a centralised facility [9]. The server running imaging software can analyse the image based on greyscale or RGB/chromaticity values, which can be correlated with the concentration of the analyte tested. The use of smartphone cameras has been proposed for diagnostic applications in dermatology [10], microscopy [11–13], ophthalmology [14], chemical analyses [15, 16] and paper-based microfluidic devices [17, 18]. However, smartphone cameras have standardisation challenges in optical analysis of colorimetric assays. For example, integrated colour balancing functions of camera phones are optimised for photography in bright ambient light. Recently, there have been several approaches to address the issues related to ambient light variability during colorimetric test readouts. For instance, a housing unit has been built to eliminate the variation in lighting conditions and positioning of the camera. These solutions required a phone-specific external housing unit and other components such as batteries, LED arrays (for reflection and transmission) and lenses [19]. In another study, a calibration chart and test assay images were captured using the phone camera, and the chromaticity diagram-based image processing was performed externally with a computer [20]. Fully-integrated smartphone apps that quantify different types of colorimetric tests on both iOS and Android platforms are needed to facilitate rapid screening at point-of-care settings.

Table 6.1 Commercial semi-quantitative urine test strips

Analyte	Test						
	Chemstrip	Multistix	Chemstrip micral	CLINITEK microalbumin	Medi-test combination	Aution sticks	URiSCAN strip
pH	✓	✓	–	–	✓	✓	✓
Glucose	✓	✓	–	–	✓	✓	✓
Protein	✓	✓	–	–	✓	✓	✓
Ketone	✓	✓	–	–	✓	✓	✓
Leukocytes	✓	✓	–	–	✓	✓	✓
Nitrite	✓	✓	–	–	✓	✓	✓
Blood	✓	✓	–	–	✓	✓	✓
Urobilinogen	✓	✓	–	–	✓	✓	✓
Bilirubin	✓	✓	–	–	✓	✓	✓
Specific gravity	✓	✓	–	–	✓	✓	✓
Ascorbic acid	–	–	–	–	✓	–	✓
Albumin	–	–	✓	✓	–	–	–
Creatinine	–	–	–	✓	–	–	–

Fig. 6.1 Mobile-cellular subscriptions in the emerging economies, developing world and worldwide from 2004 to 2014

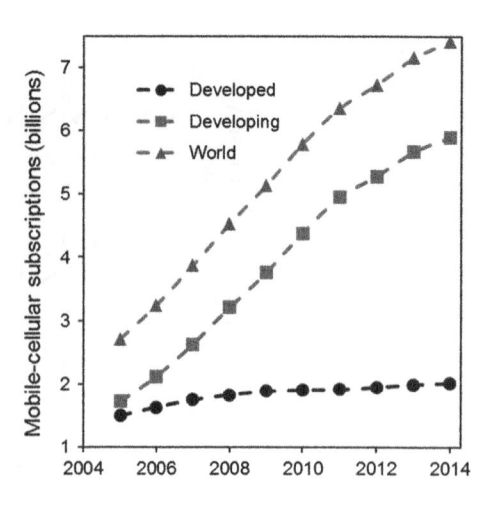

6.2 A Smartphone Algorithm for the Quantification of Colorimetric Assays

6.2.1 Calibration of the Application

The smartphone app measured the electromagnetic radiation from the colorimetric test zones with the complementary metal-oxide-semiconductor (CMOS) sensor present in the smartphone camera. The algorithm processed the colorimetric data as concentrations of the analytes in each test zone, and then the app displayed the corresponding value on the smartphone screen. The sensitivity of the measurement was based on the accuracy of the CMOS sensor, the colour uniformity of the reactions, and the number of calibration points. In testing a colorimetric urine test strip (Fig. 6.2a), the app stored a calibration curve for the assay and the ambient light condition. The user inputted (i) the sensor type, (ii) target analytes, (iii) units of the concentration, (iv) number of reference data points, and (v) the images of the calibration points were captured. The smartphone was perpendicularly positioned over the assay at 5 cm, which was kept constant to match the colorimetric zones with the evaluation area defined by the software (Fig. 6.2b). The measurements were carried out at room temperature (24 °C). Calibration was performed within ∼ 1 min, and it was stored in the smartphone memory.

The app was calibrated for pH, glucose and protein measurements based on 5, 4 and 5 data points, respectively. The app located the reference colours (100 pixels), transformed and averaged the CMOS data into non-linear red, green, blue (RGB) values (R_c, G_c, B_c) for each pixel. Subsequently, the app linearised the RGB values (R_l, G_l, B_l) [22]:

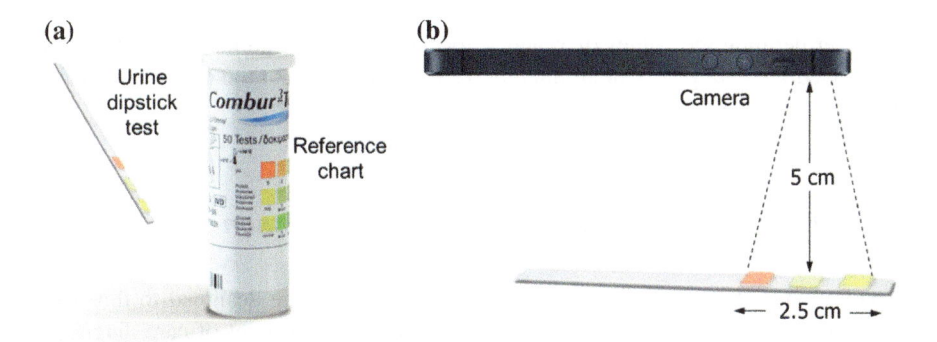

Fig. 6.2 Quantifying colorimetric tests through a smartphone reader. **a** A commercial dipstick widely used for measuring concentrations of pH, protein and glucose in urine samples. These semi-quantitative tests are low-cost and easy-to-use. Their interpretation is based on visual inspection of the developed reaction patches and comparison to a colour reference chart. **b** The smartphone captured and processed the image of the test zones, reducing time and errors related to subjective interpretation. Reprinted from Ref. [21], Copyright (2014), with permission from Elsevier

$$R_l = \left(\frac{0.055 + R_c}{1.055}\right)^{2.4} \tag{6.1}$$

$$G_l = \left(\frac{0.055 + G_c}{1.055}\right)^{2.4} \tag{6.2}$$

$$B_l = \left(\frac{0.055 + B_c}{1.055}\right)^{2.4} \tag{6.3}$$

Linear RGB values were converted to tristimulus values, X, Y, Z [22]:

$$X = 0.1805B_l + 0.3576G_l + 0.4124R_l \tag{6.4}$$

$$Y = 0.0722B_l + 0.7152G_l + 0.2126R_l \tag{6.5}$$

$$Z = 0.9505B_l + 0.1192G_l + 0.0193R_l \tag{6.6}$$

Finally X, Y, Z tristimulus values were converted to the 2D (x, y) International Commission on Illumination (CIE) 1931 chromaticity space:

$$x_j = \frac{X}{X + Y + Z} \tag{6.7}$$

$$y_j = \frac{Y}{X + Y + Z} \tag{6.8}$$

After defining the values of x_j and y_j for the jth concentration data point C_j, the app saved the data points in an internal database to complete the calibration before returning to the main screen.

6.2.2 User Interface of the Smartphone Application

The iOS (Fig. 6.3a–e) and Android (Fig. 6.3f–j) apps were designed to direct the user in all the steps necessary to perform an analysis of a point-of-care diagnostic test. First, the user might choose functions from sample testing, system calibration, sensor type selection and test history viewing (Fig. 6.3a, f). Before sample testing, the user first calibrated the system by choosing 'calibration'; this allowed capturing and processing a set of reference images of the required concentration range. The calibration was recorded under a given ambient light condition. When the ambient light changed (e.g. colour, intensity and tone), the reader needed to be recalibrated. For sample testing, the user selected 'sensor type' function to specify the type of

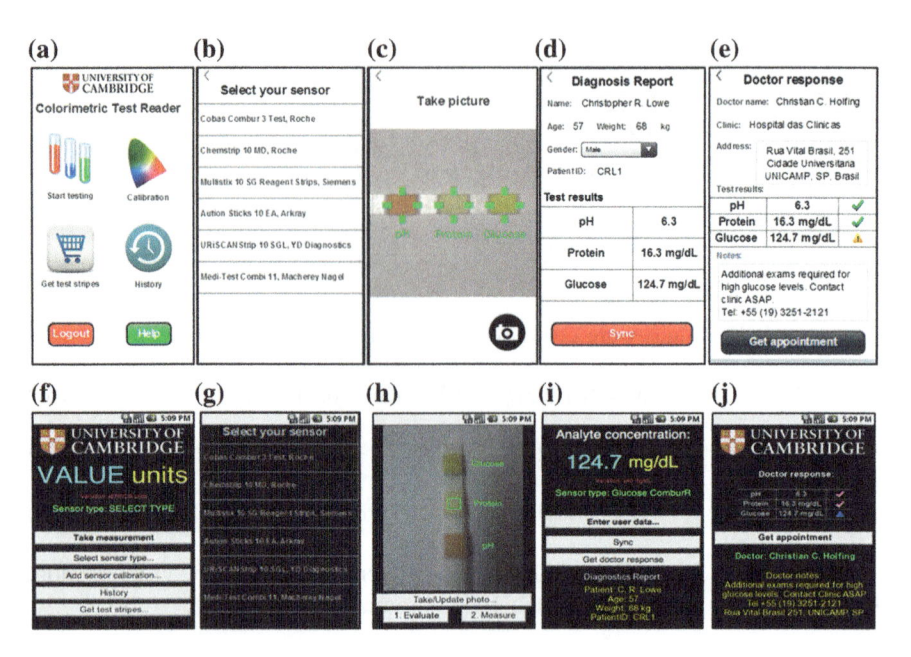

Fig. 6.3 Screenshots of the iOS (iPhone 5, 8 MP camera) and Android (Samsung I5500 Galaxy 5, 2 MP camera) apps, which quantify the colorimetric test strip for urinalysis. **a** Main menu displaying sample testing, calibration and test history viewing. **b** The app displays sensor types. **c** User captures the image of the test zones. **d** The diagnostic test results are displayed to the user. **e** The information received from the physician displayed on the final screen include user-specific instructions. **f–j** Android version of the same steps **a–e**, respectively. Reprinted from Ref. [21], Copyright (2014), with permission from Elsevier

colorimetric test to be performed (Fig. 6.3b, g). Once the sensor type was selected, the user captured the image of the corresponding test zones (Fig. 6.3c, h). The app processed the image information and transformed it into analyte concentrations by comparing the measured value with the calibration curve. Then, it displayed the results in the subject's report (Fig. 6.3d, i). This report contained personal data (including, but not limited to name, age, weight, patient ID number and contact phone number) along with the medical information that might be uploaded, synced and transmitted to an offsite doctor's office. Following review, the doctor might resend the report and any additional comments back either to the patient or to a referring practitioner (Fig. 6.3e, j).

6.2.3 Colorimetric Measurements

The app was tested on both iPhone 5 (8 MP camera) and Samsung I5500 Galaxy 5 (2 MP camera) using a colorimetric urine test strip (Cobas® Combur^3 Test®, Roche). This three-patch test strip is normally used for semi-quantitative determination of pH, protein and glucose in urine samples. Table 6.2 shows the sensing principles and limitations of the assay [23, 24]. Artificial urine samples were prepared as described previously [25]. The test strips were submerged into a range of artificial urine solutions, and the images of the test zones were analysed with the smartphone app. The calibration was performed using the colour reference chart on the product package provided by the supplier. The user captured the image of the target assay using the same conditions as for the calibration points (e.g. distance, lighting and temperature), and the app followed the same steps performed for the calibration. The images were taken at a fixed distance under an ambient fluorescent light source, whose output was measured as 10 ± 1 μW using a powermeter. The app converted the CMOS data to RGB, which were linearised, converted to tristimulus values that were expressed as measured 2D chromaticity values (x_m, y_m). Figure 6.4 shows the 2D chromaticity chart used by the app to store the calibration curve and to calculate the concentration of the target analytes.

The app computed the final measurement by comparing the target data values with respect to the calibration curve by an interpolation algorithm analogous to the nearest neighbour problem in computational geometry. For each point in the calibration curve (j), the shortest distance from the measurement value to the calibration point was determined:

$$d_k = \sqrt{(x_k - x)^2 + (y_k - y)^2} \tag{6.9}$$

where k is an integer and goes from 1 to the number of stored x and y pairs (points) in the calibration curve, i.e. $k = j$. The algorithm stored two shortest distances to the sample point: d_{ks} and d_{kss}. Their x and y values obtained by Eqs 6.7–6.8, together with their concentrations C were stored in the app memory. The concentration range, d_c, of the nearest two data points:

Table 6.2 The principles of colorimetric reactions in Cobas® Combur³ Test®, Roche

Assay	Test principle	Reactive ingredient (per 1 cm² patch area)	Range	Detection limit	Operating temp. (°C)
pH	Ingredients specifically react with H⁺; the pH is the negative common logarithm of the H_3O^+ concentration. The tests pad change colour from orange to greenish blue as the pH increase	Bromothymol blue (13.9 μg), methyl red (1.2 μg) and phenol-phthalein (8.6 μg)	5.0–9.0	0.5	stable up to boiling point
Protein	Based on protein error principle of a pH indicator. The test pad changes colour from yellow to greenish blue in the presence of albumin	3',3'',5',5''-Tetrachlorphenol-3,4,5,6-tetrabrom-sulfophthalein (neutral form) (13.9 μg)	0–100 mg/dL	6 mg albumin/dL	stable up to boiling point
Glucose	Based on glucose oxidase/peroxidase reaction. The reaction pad changes colour from yellow to dark blue in the presence of glucose	Tetramethylbenzidine (103.5 μg), glucose oxidase (6 U) and peroxidase (35 U)	0–300 mg/dL	40 mg/dL	<55

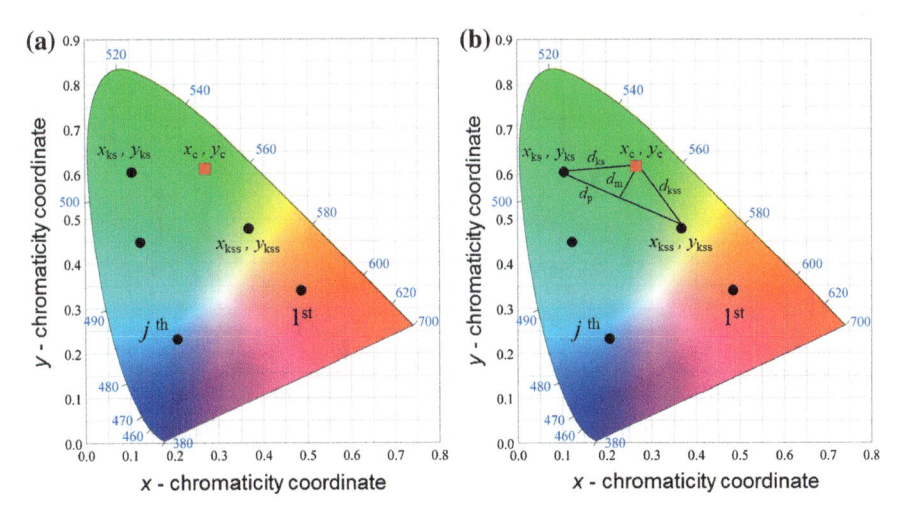

Fig. 6.4 CIE 1931 (x, y) chromaticity diagram. Monochromatic colours are located near the perimeter and *white light* is located at the centre of the diagram. **a** *Red square box* shows the location of an unknown data point. **b** The use of the near neighbour problem to obtain the position of a data point based on the calibration

$$d_C = |C_{ks} - C_{kss}| \tag{6.10}$$

where C_{ks} and C_{kss} are the concentrations of the points corresponding to d_{ks} and d_{kss}. The distance in x and y coordinates on the chromaticity space between the two nearest points on the calibration curve to the measurement point:

$$d_{xy} = \sqrt{(x_{ks} - x_{kss})^2 + (y_{ks} - y_{kss})^2} \tag{6.11}$$

The app then calculated the shortest distance from that measurement point to the line between the two calibration points:

$$d_{sd} = \frac{|(x_{kss} - x_{ks})(y_{ks} - y_m) - (x_{kss} - x_m)(y_{ks} - y_{kss})|}{\sqrt{(x_{kss} - x_{ks})^2 + (y_{kss} - y_{ks})^2}} \tag{6.12}$$

The largest among d_{ks}, d_{kss} or d_{sd} was determined and stored. The $\pm$ variation, $v_{\pm}$, was calculated as a ratio or proportion, given the concentration range d_c for the largest distance as:

$$v_{\pm} = \frac{d_m d_C}{d_{xy}} \tag{6.13}$$

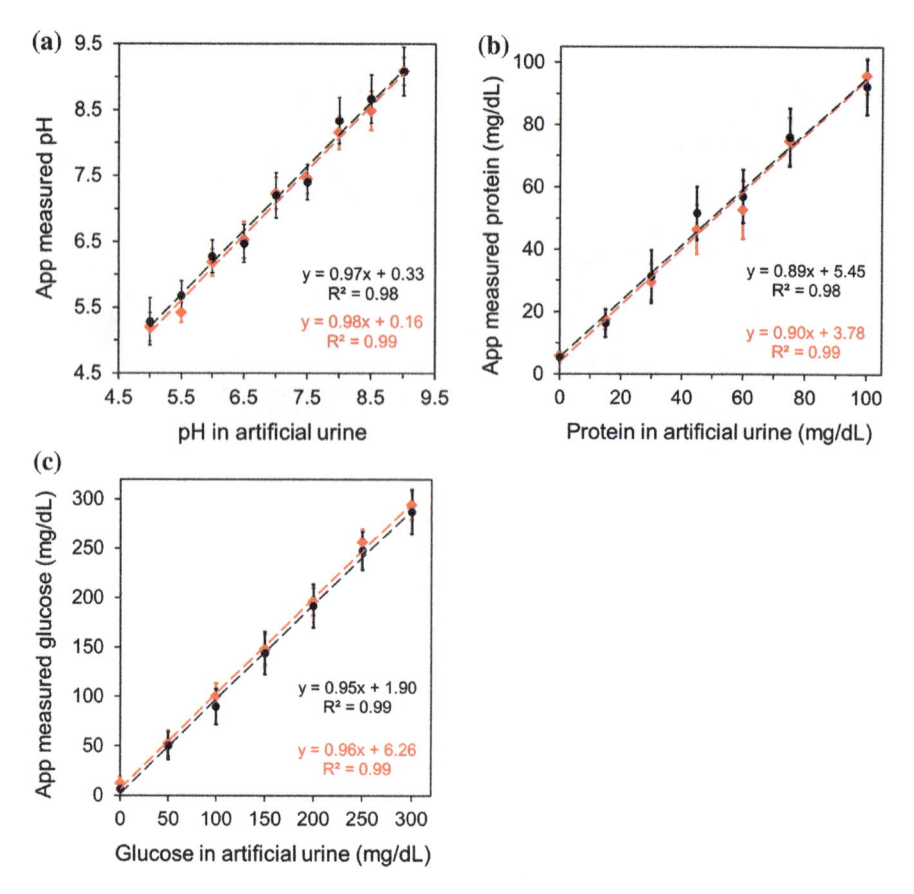

Fig. 6.5 Analyses of dipstick tests (cobas® Combur3 Test®, Roche) by iPhone 5 (shown in *red*) and Samsung I5500 Galaxy 5 (shown in black)for **a** pH, **b** protein and **c** glucose measurements in artificial urine. Standard *error bars* represent three replicates. Reprinted from Ref. [21], Copyright (2014), with permission from Elsevier

where d_m is the largest among d_{ks}, d_{kss} and d_{sd}. By using a similar proportionality approach, the distance from one of the corresponding calibration points to the point on the line to where d_{sd} was measured, was computed as the concentration (C_m):

$$C_m = \frac{d_p \, d_C}{d_{xy}} \tag{6.14}$$

$$d_p = \sqrt{d_{ks}^2 - d_{sd}^2} \tag{6.15}$$

C_m was located within the concentration range previously calculated, but it varied based on the distance between the measurement point and the calibration curve. After the algorithm ended, the app displayed the diagnostic results for the

Table 6.3 Standard deviations of residuals (s_y), slope (s_m), intercept (s_b) and limit of detection for the pH, protein and glucose measurements

Assay	Smartphone	Std. of residuals (s_y)	Std. of slope (s_m)	Std. of intercept (s_b)	Limit of detection
pH	Android	0.459	0.068	0.0161	1.66
	iPhone	0.311	0.046	0.0048	1.21
Protein (mg/dL)	Android	11.44	0.077	12.21	41
	iPhone	8.95	0.060	7.51	33
Glucose (mg/dL)	Android	27.48	0.060	53.69	92
	iPhone	20.19	0.044	20.47	69

analyte concentrations. Figure 6.5 illustrates standard curves associated with pH (5.0–9.0), protein (0–100 mg/dL) and glucose (0–300 mg/dL) measurements. These concentration values were within the physiological range. The variation in the distance, if any, was negligible as corroborated by the error in the measurements. Standard deviations of residuals (s_y), slope (s_m), intercept (s_b), and limit of detection for the pH, protein and glucose measurements are shown in Table 6.3. The app displayed accurate readouts of the respective analyte concentrations in artificial urine, demonstrating the app's potential for reading colorimetric assays.

The sensitivity of the colorimetric measurement depends on the resolution of the camera's CMOS sensor, colour uniformity of the assay reactions and the number of calibration points. The CMOS sensor registers the amount of light (electromagnetic radiation) from bright to dark without colour information, and assigns R, G, B values for each pixel. There are 256 (0–255) integer values per colour channel (R, G, B) yielding 16,777,216 different colours that can be captured by the smartphone camera. The app converts the RGB data, which can be represented in a 2D x,y plot on a chromaticity diagram (CIE 1931) with a total of 65,537 different colour values. However, the colorimetric readouts are based on a calibration in the chromaticity diagram. In terms of electromagnetic radiation received by the camera's CMOS sensor, the spectral range of a typical CMOS sensor is between ∼350 and 1,150 nm, which can be transformed to RGB and 2D x,y values. Theoretically, for each of the total different colour values in a 2D x,y plot, there is a corresponding ∼2.2 nm (i.e. 800 nm/∼362) per colour value (maximum sensitivity). More accurate estimations can be obtained by calculating the length of the parabolic calibration curve.

6.3 Conclusions

Smartphones have the potential to serve as point-of-care readers for colorimetric assays. This chapter demonstrated an algorithm that allows the smartphone camera to read semi-quantitative tests rapidly with inter-phone repeatability and minimal operator intervention. The app utilised the smartphone hardware to quantify pH and the concentrations of protein and glucose. The method of quantification was reproducible and sensitive in Android and iOS platforms. The technology has utility in resource-limited settings, where trained healthcare professionals are scarce. While this chapter demonstrated an equipment-free smartphone reader under controlled conditions, advanced apps need to be developed to automatically compensate for measurement variability due to changes in focus, angle, lighting conditions, shadow effects and sensor type. Additionally, a step forward in this process is the design of algorithms that process data efficiently into actionable information for the user. The colour conversion method demonstrated in this chapter is not limited to urinalysis, but is also applicable to other colorimetric assays such as colloidal gold, latex labels, aqueous tests, as well as emerging technologies such as holographic sensors [26–32], optical devices [33–35], plastic/PDMS-based microfluidic devices [36–42], multiplex paper- and nitrocellulose-based microfluidic devices [43, 44]. Furthermore, cloud computing can be adapted for transferring the medically-relevant data to a centralised facility, and it may be used for endemic or pandemic surveillance. The app will facilitate less expensive laboratory testing in the developed nations and enable automated readouts of point-of-care diagnostics in resource-limited settings. However, successful commercialisation of mobile medical applications will require clinical trials and FDA clearance to ensure patient safety [45].

References

1. Urdea M, Penny LA, Olmsted SS, Giovanni MY, Kaspar P, Shepherd A, Wilson P, Dahl CA, Buchsbaum S, Moeller G, Hay Burgess DC (2006) Requirements for high impact diagnostics in the developing world. Nature 444(Suppl 1):73–79. doi:10.1038/nature05448
2. Whitesides GM (2013) A glimpse into the future of diagnostics. Clin Chem 59(4):589–591. doi:10.1373/clinchem.2013.204347
3. Girosi F, Olmsted SS, Keeler E, Hay Burgess DC, Lim YW, Aledort JE, Rafael ME, Ricci KA, Boer R, Hilborne L, Derose KP, Shea MV, Beighley CM, Dahl CA, Wasserman J (2006) Developing and interpreting models to improve diagnostics in developing countries. Nature 444(Suppl 1):3–8. doi:10.1038/nature05441
4. Gordon J, Michel G (2012) Discerning trends in multiplex immunoassay technology with potential for resource-limited settings. Clin Chem 58(4):690–698. doi:10.1373/clinchem.2011.176503
5. Yager P, Domingo GJ, Gerdes J (2008) Point-of-care diagnostics for global health. Annu Rev Biomed Eng 10:107–144. doi:10.1146/annurev.bioeng.10.061807.160524
6. Gubala V, Harris LF, Ricco AJ, Tan MX, Williams DE (2012) Point of care diagnostics: status and future. Anal Chem 84(2):487–515. doi:10.1021/ac2030199
7. Webster M, Kumar V (2012) Automated doctors: cell phones as diagnostic tools. Clin Chem 58(11):1607–1609. doi:10.1373/clinchem.2012.194555
8. Mobile-Cellular Subscriptions (2014) International Telecommunication Union, Place des Nations. http://www.itu.int. Accessed 27 Oct 2014
9. Lee DS, Jeon BG, Ihm C, Park JK, Jung MY (2011) A simple and smart telemedicine device for developing regions: a pocket-sized colorimetric reader. Lab Chip 11(1):120–126. doi:10.1039/c0lc00209g

10. Kroemer S, Fruhauf J, Campbell TM, Massone C, Schwantzer G, Soyer HP, Hofmann-Wellenhof R (2011) Mobile teledermatology for skin tumour screening: diagnostic accuracy of clinical and dermoscopic image tele-evaluation using cellular phones. Br J Dermatol 164 (5):973–979. doi:10.1111/j.1365-2133.2011.10208.x

11. Breslauer DN, Maamari RN, Switz NA, Lam WA, Fletcher DA (2009) Mobile phone based clinical microscopy for global health applications. PLoS ONE 4(7):e6320. doi:10.1371/journal.pone.0006320

12. Tseng D, Mudanyali O, Oztoprak C, Isikman SO, Sencan I, Yaglidere O, Ozcan A (2010) Lensfree microscopy on a cellphone. Lab Chip 10(14):1787–1792. doi:10.1039/c003477k

13. Smith ZJ, Chu K, Espenson AR, Rahimzadeh M, Gryshuk A, Molinaro M, Dwyre DM, Lane S, Matthews D, Wachsmann-Hogiu S (2011) Cell-phone-based platform for biomedical device development and education applications. PLoS ONE 6(3):e17150. doi:10.1371/journal.pone.0017150

14. Pamplona VF, Mohan A, Oliveira MM, Raskar R (2010) Dual of shack-hartmann optometry using mobile phones, frontiers in optics. In: OSA technical digest (CD), Optical Society of America, p FTuB4. doi:10.1364/FIO.2010.FTuB4

15. Zhu H, Mavandadi S, Coskun AF, Yaglidere O, Ozcan A (2011) Optofluidic fluorescent imaging cytometry on a cell phone. Anal Chem 83(17):6641–6647. doi:10.1021/ac201587a

16. Coskun A, Wong J, Khodadadi D, Nagi R, Tey A, Ozcan A (2012) A personalized food allergen testing platform on a cellphone. Lab Chip 13:636–640. doi:10.1039/C2LC41152K

17. Martinez AW, Phillips ST, Whitesides GM (2008) Three-dimensional microfluidic devices fabricated in layered paper and tape. Proc Natl Acad Sci USA 105(50):19606–19611. doi:10.1073/pnas.0810903105

18. Pollock NR, Rolland JP, Kumar S, Beattie PD, Jain S, Noubary F, Wong VL, Pohlmann RA, Ryan US, Whitesides GM (2012) A paper-based multiplexed transaminase test for low-cost, point-of-care liver function testing. Sci Transl Med 4 (152):152ra129. doi:10.1126/scitranslmed.3003981

19. Mudanyali O, Dimitrov S, Sikora U, Padmanabhan S, Navruz I, Ozcan A (2012) Integrated rapid-diagnostic-test reader platform on a cellphone. Lab Chip 12(15):2678–2686. doi:10.1039/c2lc40235a

20. Shen L, Hagen JA, Papautsky I (2012) Point-of-care colorimetric detection with a smartphone. Lab Chip 12(21):4240–4243. doi:10.1039/c2lc40741h

21. Yetisen AK, Martinez-Hurtado JL, Garcia-Melendrez A, Vasconcellos FC, Lowe CR (2014) A smartphone algorithm with inter-phone repeatability for the analysis of colorimetric tests. Sens Actuators B 196:156–160. doi:10.1016/j.snb.2014.01.077

22. Hunt RW (1998) Measuring colour, 3rd edn. Fountain Press, London

23. Gouda MD, Singh SA, Rao AG, Thakur MS, Karanth NG (2003) Thermal inactivation of glucose oxidase. Mechanism and stabilization using additives. J Biol Chem 278 (27):24324–24333. doi:10.1074/jbc.M208711200

24. Zoldak G, Zubrik A, Musatov A, Stupak M, Sedlak E (2004) Irreversible thermal denaturation of glucose oxidase from Aspergillus niger is the transition to the denatured state with residual structure. J Biol Chem 279(46):47601–47609. doi:10.1074/jbc.M406883200

25. Brooks T, Keevil CW (1997) A simple artificial urine for the growth of urinary pathogens. Lett Appl Microbiol 24(3):203–206. doi:10.1046/j.1472-765X.1997.00378.x

26. Yetisen AK, Montelongo Y, da Cruz Vasconcellos F, Martinez-Hurtado JL, Neupane S, Butt H, Qasim MM, Blyth J, Burling K, Carmody JB, Evans M, Wilkinson TD, Kubota LT, Monteiro MJ, Lowe CR (2014) Reusable, robust, and accurate laser-generated photonic nanosensor. Nano Lett 14(6):3587–3593. doi:10.1021/nl5012504

27. Yetisen AK, Qasim MM, Nosheen S, Wilkinson TD, Lowe CR (2014) Pulsed laser writing of holographic nanosensors. J Mater Chem C 2(18):3569–3576. doi:10.1039/C3tc32507e

28. Yetisen AK, Butt H, da Cruz Vasconcellos F, Montelongo Y, Davidson CAB, Blyth J, Chan L, Carmody JB, Vignolini S, Steiner U, Baumberg JJ, Wilkinson TD, Lowe CR (2014) Light-directed writing of chemically tunable narrow-band holographic sensors. Adv Opt Mater 2 (3):250–254. doi:10.1002/adom.201300375

29. Yetisen AK, Naydenova I, Vasconcellos FC, Blyth J, Lowe CR (2014) Holographic sensors: three-dimensional analyte-sensitive nanostructures and their applications. Chem Rev 114 (20):10654–10696. doi:10.1021/cr500116a

30. Yetisen AK, Montelongo Y, Qasim MM, Butt H, Wilkinson TD, Monteiro MJ, Lowe CR, Yun SH (2014) Nanocrystal bragg grating sensor for colorimetric detection of metal ions (under review)

31. Tsangarides CP, Yetisen AK, da Cruz Vasconcellos F, Montelongo Y, Qasim MM, Wilkinson TD, Lowe CR, Butt H (2014) Computational modelling and characterisation of nanoparticle-based tuneable photonic crystal sensors. RSC Adv 4(21):10454–10461. doi:10.1039/C3RA47984F

32. Farandos NM, Yetisen AK, Monteiro MJ, Lowe CR, Yun SH (2014) Contact lens sensors in ocular diagnostics. Adv Healthc Mater. doi:10.1002/adhm.201400504

33. Deng S, Yetisen AK, Jiang K, Butt H (2014) Computational modelling of a graphene Fresnel lens on different substrates. RSC Adv 4(57):30050–30058. doi:10.1039/C4ra03991b

34. Kong X-T, Butt H, Yetisen AK, Kangwanwatana C, Montelongo Y, Deng S, Fd Cruz Vasconcellos, Qasim MM, Wilkinson TD, Dai Q (2014) Enhanced reflection from inverse tapered nanocone arrays. Appl Phys Lett 105(5):053108. doi:10.1063/1.4892580

35. Vasconcellos FD, Yetisen AK, Montelongo Y, Butt H, Grigore A, Davidson CAB, Blyth J, Monteiro MJ, Wilkinson TD, Lowe CR (2014) Printable surface holograms via laser ablation. ACS Photonics 1(6):489–495. doi:10.1021/Ph400149m

36. Yager P, Edwards T, Fu E, Helton K, Nelson K, Tam MR, Weigl BH (2006) Microfluidic diagnostic technologies for global public health. Nature 442(7101):412–418. doi:10.1038/nature05064

37. Chin CD, Laksanasopin T, Cheung YK, Steinmiller D, Linder V, Parsa H, Wang J, Moore H, Rouse R, Umviligihozo G, Karita E, Mwambarangwe L, Braunstein SL, van de Wijgert J, Sahabo R, Justman JE, El-Sadr W, Sia SK (2011) Microfluidics-based diagnostics of infectious diseases in the developing world. Nat Med 17(8):1015–1019. doi:10.1038/nm.2408

38. Chin CD, Cheung YK, Laksanasopin T, Modena MM, Chin SY, Sridhara AA, Steinmiller D, Linder V, Mushingantahe J, Umviligihozo G, Karita E, Mwambarangwe L, Braunstein SL, van de Wijgert J, Sahabo R, Justman JE, El-Sadr W, Sia SK (2013) Mobile device for disease diagnosis and data tracking in resource-limited settings. Clin Chem 59(4):629–640. doi:10.1373/clinchem.2012.199596

39. Yang X, Piety NZ, Vignes SM, Benton MS, Kanter J, Shevkoplyas SS (2013) Simple paper-based test for measuring blood hemoglobin concentration in resource-limited settings. Clin Chem 59(10):1506–1513. doi:10.1373/clinchem.2013.204701

40. Yetisen AK, Volpatti LR (2014) Patent protection and licensing in microfluidics. Lab Chip 14 (13):2217–2225. doi:10.1039/c4lc00399c

41. Volpatti LR, Yetisen AK (2014) Commercialization of microfluidic devices. Trends Biotechnol 32(7):347–350. doi:10.1016/j.tibtech.2014.04.010

42. Yetisen AK, Jiang L, Cooper JR, Qin Y, Palanivelu R, Zohar Y (2011) A microsystem-based assay for studying pollen tube guidance in plant reproduction. J Micromech Microeng 21 (5):054018. doi:10.1088/0960-1317/21/5/054018

43. Yetisen AK, Akram MS, Lowe CR (2013) Paper-based microfluidic point-of-care diagnostic devices. Lab Chip 13(12):2210–2251. doi:10.1039/c3lc50169h

44. Akram MS, Daly R, Vasconcellos FC, Yetisen AK, Hutchings I, Hall EAH (2015) Applications of paper-based diagnostics. In: Castillo-Leon J, Svendsen WE (eds) Lab-on-a-Chip Devices and Micro-Total Analysis Systems. Springer, Berlin

45. Yetisen AK, Martinez-Hurtado JL, da Cruz Vasconcellos F, Simsekler MC, Akram MS, Lowe CR (2014) The regulation of mobile medical applications. Lab Chip 14(5):833–840. doi:10.1039/c3lc51235e

Chapter 7
The Prospects for Holographic Sensors

The development of rapid and low-cost optical sensors can enable monitoring of high-risk individuals at point of care. This thesis described the design, fabrication and optimisation of holographic pH [1, 2], divalent metal cation [3], and glucose sensors [4]. Holographic sensing is an emerging analytical platform that allows semi-quantitative colorimetric readouts by eye and fully-quantitative results by spectrophotometry. They have the added advantage of being rapidly fabricated using laser light and having precise control over the optical characteristics as compared to other optical sensors. This chapter discusses potential areas of research in (i) fabricating holographic sensors, (ii) functionalising the hydrogel matrices to increase the capabilities and the performance, (iii) multiplexing holographic sensors through microfluidics, and (iv) extracting quantitative readouts via smartphone and wearable devices. Additionally, this chapter identifies the gaps within the field, outlines the strategies to overcome the perceived limitations of holographic sensors, and includes challenges to scaling up and commercialisation.

7.1 The Development of Fabrication Approaches

To design analytical devices with predictive optical characteristics and optimise the parameters that influence the sensing performance, the holographic sensors were evaluated using computational simulations through finite element modelling [2]. The simulated sensor consisted of a multilayer structure with alternating refractive indices. The model allowed the analyses of optical properties on varying the pattern and characteristics of the Ag^0 nanoparticle (NP) arrays within a theoretical hydrogel matrix. Factors such as variation in Ag^0 NP diameter and distribution within the hydrogel matrices were computationally studied to analyse the properties of reflected and absorbed light. The geometry of the photonic structure can be modified to obtain optical sensors with desired degree of diffraction angle and bandwidth. The simulations showed that the intensity of the reflection band increased as the density of Ag^0 NPs increased within the polymer matrix up to a point, where insufficient light was able to reach all the multilayer gratings. In addition, the model also showed that the

© Springer International Publishing Switzerland 2015
A.K. Yetisen, *Holographic Sensors*, Springer Theses,
DOI 10.1007/978-3-319-13584-7_7

density of Ag^0 NP was correlated with both the depth and width of the spectra. It was feasible to use commercial simulation software as compared to the analytical solution since, the device geometry consisted of Ag^0 NPs with different sizes. COMSOL Multiphysics® allowed having control over the design of the multilayer structure with easily varying the size and spatial distribution of Ag^0 NPs. Understanding the design parameters and their relation to optical characteristics of the proposed sensor allowed the rational design of sensors with control over the entire mechanism. This has advantages in estimating the diameter and distribution of Ag^0 NPs and avoiding wider band gaps to obtain precise and tuneable optical devices. Many further applications can follow from these simulations ranging from photonic structures, where specific optical properties can lead to unique light-diffraction spectra. A limitation of the finite element modelling in this thesis was that reflection (or absorption) simulations were performed at small geometries due to limitations in the computational memory. For accurate analysis, diffraction profile should be simulated at large geometries.

The sensors were fabricated through silver-halide chemistry [2], in situ size reduction of Ag^0 NPs [1, 4], and photopolymerisation [3] to form holographic poly (2-hydroxyethyl methacrylate) (pHEMA) and polyacrylamide (pAAm) matrices. A technique was developed to deposit recording media on poly(methyl methacrylate) (PMMA) substrates. This involved O_2 plasma treatment of PMMA to render its surface hydrophilic. In holographic pH sensors, a monomer mixture consisting of 2-hydroxyethyl methacrylate (HEMA), ethylene dimethacrylate (EDMA) and methacrylic acid (MAA) were UV-initiated free-radical copolymerised using dimethoxy-2-phenylacetophenone (DMPA) on PMMA substrates. In holographic glucose sensors, the polymeric backbone was formed from acrylamide, N, N'-methylenebisacrylamide (MBAAm) and 3-(acrylamido)phenylboronic acid (3-APB), which were copolymerised on PMMA substrates. The next step in the holographic sensor fabrication involved doping hydrogel matrices with Ag^0 NPs ($\emptyset$ 10–100 nm). This was achieved by perfusing silver ions (Ag^+) into the hydrogel matrices, and subsequently reducing Ag^+ ions to Ag^0 NPs using a photographic developer, which acted as a reducing agent. After the recording medium was prepared, the diffraction gratings were formed within the hydrogel matrices, which reported on the concentration of target analytes. For creating Bragg gratings, holography was chosen due to its low cost and amenability to mass manufacturing. A single pulse of a laser (6 ns, 532 nm, 350 mJ) in "Denisyuk" reflection mode formed diffraction gratings consisting of ordered Ag^0 NPs embedded within the hydrogel matrices [1, 4]. This laser writing involved the use of the standing wave, where the energy at the antinodes reduced the size distribution of Ag^0 NPs from ~ 10–100 to ~ 15–30 nm. Upon illumination with a while light source, the fabricated photonic structure diffracted monochromatic (narrow-band) light. The optical characteristics of the diffracted light depended on the wavelength of the laser light, which was used to produce the grating. The novelty of the platform concerned the use of a single 6 ns highly-intense laser pulse (350 mJ) to reduce directly the size distribution of Ag^0 NPs at the constructive interference sites (antinodes) in hydrogel matrices. This created ablated and non-ablated planes within the hydrogel matrix, producing a refractive index contrast in the form of a multilayer diffraction grating.

This method was fundamentally different from silver halide chemistry, in which the latent image (small cluster of Ag^0 atoms) on photosensitive silver bromide (AgBr) nanocrystals was photographically developed to form a multilayer diffraction grating [5]. In TEM images, there was a distinction in the reduced size of Ag^0 NPs in hydrogel matrices before and after photochemical patterning. Based on the angular-resolved measurements, it was inferred that the photochemical patterning formed Bragg diffraction gratings as well as a blazed grating (tooth saw) or a transmission grating, as predicted by the simulations based on the superposition of waves [1]. While the optical microscopic images of the surface gratings supported the simulated results, no direct evidence of the multilayer structure was found in SEM, ESEM and TEM images. This could be attributed to a low density of Ag^0 NPs within the hydrogel matrices, which might be due to the low (<1 %) diffraction efficiency of the photonic structure. However, the period (~ 3 μm) inferred from the transmission gratings indirectly supported the existence of the multilayer structure [6, 7].

An advantage of holographic sensors fabricated through in situ size reduction of Ag^0 NPs is that this method does not require a gentle reduction step after laser exposure. This allows the use of any reducing agent, not limiting the reduction process to weak developers. For example, non-aqueous agents (e.g. hydroquinone in THF) can be used, and tedious control over exposure parameters and Ag^0 NP growth is not needed. The present technique shows that Bragg diffraction gratings can be produced quickly with reduced complexity as compared to silver-halide chemistry-based fabrication. This opens holographic sensors into a broader range of applications and materials. For example, gold, copper or iron NPs can be used to produce the multilayer gratings. A limitation of the present photochemical patterning of Ag^0 NPs is that the photonic structure is not compatible with bleaching with bromine (Br_2), chlorine (Cl_2), and iodine (I_2). This may be attributed to the uncontrolled growth of AgBr nanocrystals at the antinode as well as the node regions. Another explanation for this phenomenon is that the Ag^0 NPs, which may be burnt in the hydrogel matrix by the high-power of the laser light, become detached from the matrix so that they loose their spatial integrity. Therefore, it may not be possible to fabricate Ag^0 NP holograms with diffraction efficiency >10 %. This is also a limitation in sensing samples with divalent cations such as Cu^{2+}, which cause a decrease in the diffraction efficiency due to bleaching. The future work may explore titanium(IV) oxide NPs, which have a refractive index of ~ 2.8 at 632.8 nm [8, 9]. These particles may be incorporated into the hydrogel matrices to form diffraction gratings. Other attractive materials and structures might include graphene [10, 11], nanocones [12], graphite and carbon nanotubes [13]. The perfusion of Ag^+ ions into hydrogel matrices and subsequent reduction in situ allows forming Ag^0 NPs in the upper half of the hydrogel matrix (~ 5–10 μm) down to 5–6 nm from the hydrogel-air surface [14]. There are two main contributors to this issue: (i) The depletion of the developer strength as it perfuses into the hydrogel matrix, and (ii) the use of Ag^+ ions dissolved in aqueous solutions, which may not allow the Ag^+ ions to diffuse throughout the matrix. Hence, methods to obtain Ag^0 NPs throughout the hydrogel matrix should be developed to increase the number of multilayer gratings, which will increase the diffraction efficiency. A plausible approach is to synthesise monomer mixtures that incorporate metal NPs in organic solvents (0.1–1 mg/mL) produced by pulsed laser ablation [15, 16].

Another fabrication approach that has been explored in this thesis was to synthesise porphyrin derivatives that function as the crosslinker, the light absorbing material, and the component of a diffraction grating. The use of this multifunctional porphyrin permitted rapid fabrication of a photonic structure, which diffracted narrow-band light. The sensor was fabricated in Denisyuk reflection mode, however, no NP formation was required. As opposed to physical size reduction of Ag^0 NPs; in photopolymerisation, the porphyrin molecule absorbs the laser light and further polymerises the hydrogel matrix at the antinode regions. This creates a refractive index contrast of alternating polymerised and highly polymerised layers. Since photopolymerisation is NP-free, it substantially reduces the requirements for fabrication and eliminates the bleaching issues [17–20]. Photopolymerisation will play greater roles in the optimisation of the performance of holographic sensors and reduce batch to batch variability.

Printing techniques or the use of photomasks during fabrication can introduce user-friendly fool-proof text/quantity-reporting capabilities. Printing of recording media is an untapped way of depositing the monomer solution and photosensitive materials on substrates. This approach has practicality, miniaturisation capacity and scalability in the deposition of holographic materials and construction of holographic arrays. For example, various polymers can be loaded on a cartridge that can be fit to noncontact, contact or airjet dispensers. As the monomers immobilise on the surface of the substrate at room temperature or through UV-initiated free-radical polymerisation, holographic arrays consisting of different analyte-sensitive materials can be constructed with controlled size. Since this approach minimises the required volume of the monomer solutions, it has the potential to enable mass production. Printing can also allow depositing holographic sensors on implantable chips and contact lens sensors [21]. Furthermore, holographic sensor fabrication is based on the use of a planar mirror, which produces an angular intolerant hologram that limits the angle of view to ±10°. Diffusers or lenses can be used to improve the angular tolerance; however, holograms produced by this method have low-diffraction efficiency (<5 %) [22]. Additionally, digital holography (no real object requirement) may increase the diffraction efficiency, and allow multiplexing through superpositioning of images [23, 24].

7.2 Ligand Chemistry

This thesis described the ligands for holographic pH [1, 2], metal cation [3] and glucose sensors [4]. In sensing pH, the pHEMA matrix incorporated carboxylic acid groups. When ionisable groups are copolymerised with other monomers, their individual functions dictate the degree of pH-sensitivity of the pHEMA matrix. The holographic pH sensors were tested in phosphate buffers and artificial urine to show their putative clinical application. The Bragg grating was finely modulated to diffract narrow-band light based on the volumetric changes of the matrix, Ag^0 NP spacing, and index of refraction. When the pendant functional groups were ionised (e.g. deprotonating carboxylic acid groups), the polymer swelled due to the electrostatic and Donnan osmotic pressure forces that drew in or expelled counterions along with water molecules, which altered the periodicity (lattice spacing) of the diffraction grating. An increase or decrease in the lattice spacing shifted the diffracted light to longer or shorter wavelengths. This systematic lattice modulation consequently allowed a quantitative readout through wavelength changes of the diffracted light, enabling spectroscopic measurement of colour changes as a function of pH. These sensors produced visually perceptible and reversible colour changes at either side of the apparent dissociation constant (pK_a) as a function of the pH. The holographic sensors diffracted narrow-band light from the visible region to the near-infrared region ($\lambda_{peak} \approx 495$–$815$ nm). The clinical application of the diffraction gratings was demonstrated by pH sensing of artificial urine over the physiological range (4.50–9.00), with a sensitivity of 48 nm/pH unit between pH 5.0 and 6.0. As the pH was increased, the Bragg peak shifted to longer wavelengths while the intensity decreased. This was attributed to a decrease in the effective refractive index contrast since the expanding structure lowered the density of Ag^0 NPs present within a given volume. This was predicted by the model; both simulations and experimental readouts showed agreement about the Bragg peak shift of the diffracted light based on the lattice expansion and contraction.

In addition to holographic sensors, several advances in pH sensing have been demonstrated. Electrochemical and field-effect transistor based pH sensors have utilised carbon fibre [25] and carbon/metallic [26] nanostructures, respectively. Recent fluorescent sensors based on Förster resonance energy transfer using synthetic DNA [27], genetically encoded red protein [28] and an antibody-conjugated pH dependent dye [29] have been employed for intracellular monitoring. These sensing mechanisms have selectivities down to 0.01 pH units from pH 2.0 to 12.0 [30]. Holographic sensors have a comparable accuracy to these sensors in the range from pH 5.0 to 6.0; however, the dynamic range of holographic sensors needs to be improved by incorporating different functional groups to induce a Bragg shift in the desired dynamic range. The range of the pH sensitivity can be tuned by selecting the desired acidic or basic monomers to cover the pH range of the application of interest. For example, other functional groups can be chosen from trifluoromethyl propenoic acid (TFMPA), dimethylaminoethyl methacrylate (DMAEM), and vinyl imidazole to achieve a pH range from 2.0 to 9.0. In comparison to other colorimetric sensors, for example, pH dependent dye-based sensors can only be used once, while holographic sensors have the capability of being used for multiple non-consumptive analyses. As opposed to electrochemical sensing, holographic sensors do not require external power to operate for visual readouts. Additionally, since the laser light forms the image of a planar mirror, the resulting photonic structure produces an unidirectional diffraction, allowing readouts up to a meter away. Dyes and fluorescence risk losing distinct signal, and remote wireless readings are not feasible.

For sensing metal cations, a porphyrin derivative was synthesised to function as the chelating agent in the hydrogel matrix. The Bragg diffraction gratings in the polymer matrix were formed through photopolymerisation. As a proof-of-concept, the new sensing platform was characterised through its reversible colorimetric tuneability in response to variation in the concentrations of organic solvents in water (i.e. 10 %, v/v, ethanol, methanol, propan-2-ol and DMSO) within the visible region of the spectrum ($\lambda_{max} \approx$ 520–680 nm) with a response time within 50 s. The sensing mechanism allowed measuring Cu^{2+} and Fe^{2+} cations from 50.0 mM to 1.0 M; however, μ/nm concentrations were not able to be detected with the sensor. This can be attributed to the porphyrin derivative, which served as the crosslinker as well as the chelating agent. Therefore, the concentration of porphyrin derivative within the hydrogel matrix was limited by the maximum amount of crosslinker that could be accommodated, which in turn reduced the elasticity of the matrix. In contrast, the synthesis of pendant porphyrin compounds can enhance the sensitivity of the holographic sensor. In addition to the porphyrin-based sensor, other chelating agents such as 8-hydroxyquinoline [31–34] and 4-acryloylamidobenzo-18-crown-6 [35] can be incorporated in pAAm matrices for sensing metal cations.

Glucose sensors were fabricated through in situ size reduction of Ag^0 NPs. Glucose was measured from 0.1 to 10.0 mM with a minimum detection limit of 90 μM. However, the sensor response required ~ 1 h to saturate due to slow binding kinetics of *cis* diols of 3-APB with glucose molecules. This was a limitation as compared to electrochemical sensors, which could provide readouts under a minute. A theoretical kinetic model was developed to correlate the speed (slope) of Bragg peak shift with the concentration of glucose in the sample. The model assumed that the sensor would reach a saturation point, and based on the slope, it assigned a value for the concentration of glucose. The same model can be applied to any hydrogel-based sensor in order to decrease the turnaround time, where the receptor-analyte (or matrix effects) binding occurs at a slow rate. The range of sensor response and the sensitivity of the hydrogel were modulated by varying the concentration of the MBAAm and 3-APB in the hydrogel matrix, respectively. The sensor response can also be accelerated by copolymerising *n*-hexylacrylate into an acrylamide-bisacrylamide hydrogel to obtain a microporous hydrogel, which may reach equilibrium under 3 min [36].

Clinical trials in the urine samples of diabetic patients have shown that the sensor could respond to glucose at concentrations up to 375 mM within 5 min with a reset time to the baseline in ~ 10 s. Measurements indicated that the holographic glucose sensor had an improved R^2 value (0.79) as compared to commercial urine test strips read by their associated readers (0.28), while showing comparable accuracy (y = 1.2x) to the fully-automated analytical instruments. Interference due to pH, lactate, fructose and osmolality were also investigated. Although the interference due to lactate and fructose were less than 3 % in the urine samples, the sensor's specificity in measuring glucose in other biological fluids such as blood poses a challenge since fructose may be present at higher concentrations. A number of strategies have been developed to increase the specificity to glucose, and in particularly reducing the interference from lactate. Phenyl boronic acid derivatives, such as 2-(acrylamido)phenylboronate (2-APB) [37], 2-acrylamido-5-fluor-ophenylboronic acid (5-F-2-MAPBA) [38], 4-vinylphenylboronic acid (4-VPBA) [39], and their combinations with other comonomers [40, 41] can be used to improve specificity to glucose.

Like other hydrogel-based optical sensors, holographic sensors suffer from low selectivity, which stands out as an issue for the development of products that can compete with molecular dyes, electrochemical sensors and lateral-flow assays. Therefore, the emphasis on the functionalisation and the optimisation of holographic recording media needs to be increased. This requires further investigations in the design of new receptors that can selectively bind to target analytes. Understanding the fundamentals of binding kinetics and reversibility will allow constructing assays with improved control for real-time continuous monitoring applications. Studies on hydrogel dynamics and characteristics including expansion, shrinkage, diffraction efficiency, control over NP size distribution, hydrogel-analyte interactions, surface energy, release characteristics, reversibility, control of hydrogel pore size, polymer decay, effect of porous and solid nanodopants, and effect of temperature and moisture will be explored in the realisation of holographic sensors. The quality and shelf life of the holographic sensor after long-term storage also require investigations.

7.3 Multiplexing Holographic Sensors with Microfluidic Devices

This thesis also described the initial steps towards using holographic sensors in the flake form and their integration with strip substrates. Paper matrices and nitrocellulose membranes were utilised as wicking substrates to produce test strips. Three techniques were developed to reduce the background noise due to reflection from the substrate: (i) Impregnating paper with Fe_3O_4, (ii) dyeing chromatography and filter papers to black with Procion Black MX-K, and (iii) dyeing nitrocellulose membranes with DEKA-L fabric dyes. The resulting strip membranes were used to wick and deliver the sample to holographic pH sensor flakes to produce visual colour changes. The test strips can be patterned by printing wax and be assembled with lamination sheets to form multiplex paper-based microfluidic assays (Fig. 7.1a–d). In constructing high-throughput devices, any colorimetric sensor can be multiplexed; however, they require physical separation, multiple dyes and fluorophores, which are (i) not directional, (ii) work in different wavelength ranges, (iii) may have cross talk either chemically or optically. Hydrogel-based sensors in the present work can be multiplexed such that these complications are avoided. Furthermore, in sensing glucose, the pH of the urine was corrected to 7.40 using an alkaline solution, which might not be feasible at point-of-care settings. Additionally, the holographic sensors can be affected from the variation in ionic strength; therefore, pH and ionic strength sensors may be multiplexed with other sensors

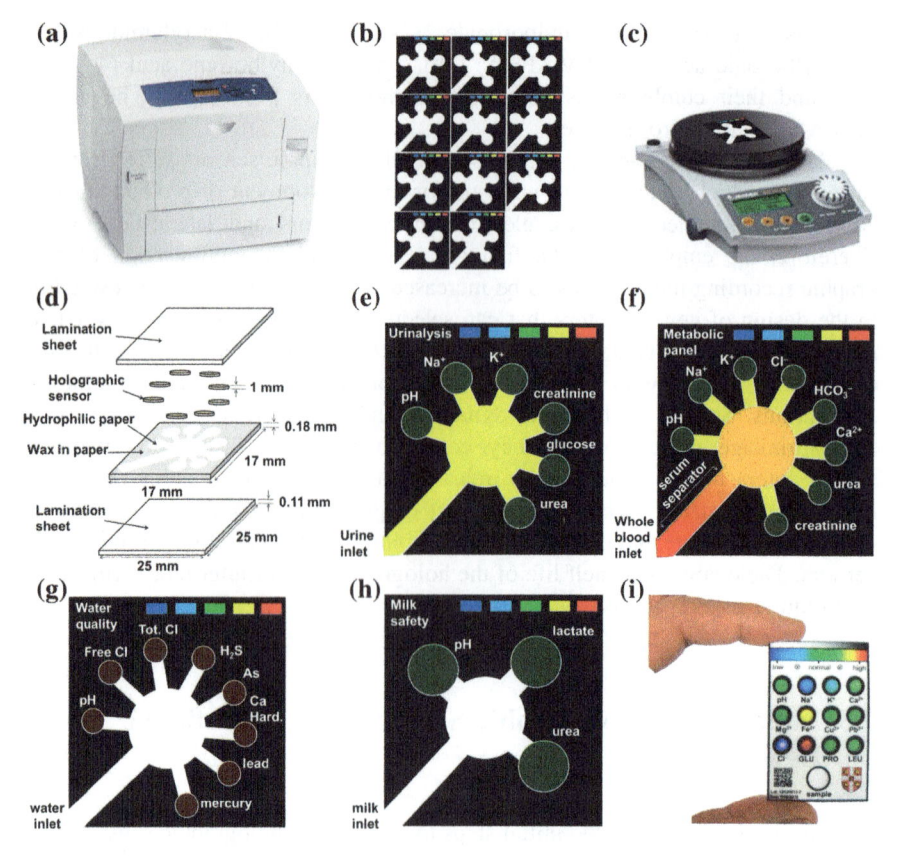

Fig. 7.1 Semi-quantitative holographic lab-on-a-chip devices. **a** Wax printing using Xerox ColorQube 8570, **b** patterned filter paper, **c** Hot plate treatment at 150 °C for 2 min. **d** Assembly of the holographic sensors on paper-based microfluidic devices. Quantification of analytes in samples of **e** urine, **f** whole blood, **g** water and **h** milk. **i** A prototype of a holographic microfluidic device

using microfluidics, and the readouts may be corrected [5, 42]. Moving forward, practical approaches require a lab-on-a-chip assay to achieve a commercial device [43–47]. Figure 7.1e–h illustrates a range of holographic sensors that can be integrated with paper-based microfluidic devices. Figure 7.1i shows a lab-on-a-chip device that allows multiplexing holographic sensors.

7.4 Readouts with Smartphones and Wearable Devices

Medical smartphone app readers can enable the decentralisation of healthcare through low-cost and highly portable point-of-care diagnostics. A generic smartphone algorithm was developed and tested for the quantification of colorimetric

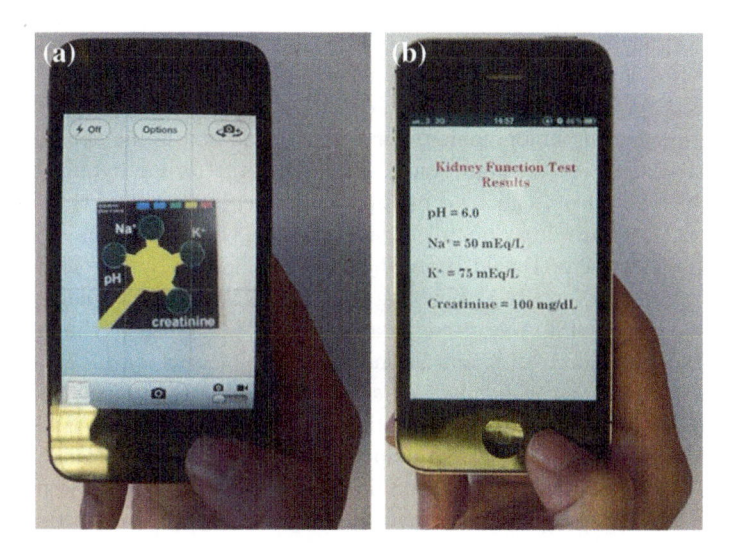

Fig. 7.2 A prototype of a holographic paper-based microfluidic device interpreted with a smartphone application. **a** Capturing image from the assay, and **b** Displaying the results

assays [48]. The app allowed automated analysis of the colorimetric tests. It utilised the nearest neighbour problem approach to estimate the pH, protein and glucose concentrations in artificial urine solutions using commercial test strips. The app used the phone's camera to convert colorimetric data from the assay into numerical concentration values on the phone's screen. It performed this by comparing the colorimetric data with a pre-recorded calibration by locating the position of the colour point data in the model and assigning it a concentration value. The app was tested with both an Android phone and an iPhone, showing comparable accuracy. The results can be stored, sent to a healthcare professional, or directly analysed by the phone for diagnosis. The app allows users to collect quantitative data and submit the results to clinicians for immediate review. By quickly getting medical data from the field to doctors or centralised laboratories can slow or limit the spread of pandemics. Future work in this area should address automated compensation for measurement variability (e.g. focus, angle, lighting conditions, shadow effects and sensor type). Advanced algorithms should eventually produce actionable information for the user while also ensuring patient safety [49]. Investigations of data processing strategies and auto-compensation capabilities to mitigate errors due to colorimetric interference will be valuable contributions to the field. Such developments may create solutions for reducing interference from the background colour of the samples (e.g. blood). Quantitative data processing on mobile devices such as smartphones, tablets, smartwatches, Google Glass should enhance the compatibility and feasibility in developing integrated holographic diagnostic devices. Figure 7.2 illustrates the prototype of an app that allows reading paper-based holographic devices. Although significant time has been devoted to spectrophotometric readouts, the attribute of being equipment-free should not be overlooked. The use of external readers is a barrier for the existing assays, yet this will be a greater challenge in adopting holographic diagnostics. In addition to the app described in this thesis, there are newer apps that might compensate for the lighting conditions [50] and location identification of multiple colorimetric sensors [51]. Direct competition of the app developed in this thesis is handheld readers. Commercial readers such as RDS-1500 Pro (Detekt Biomedical, Austin, TX), Defender TSR™ R-5001 (Alexeter Technologies LLC, Wheeling, IL), LFDR101 (Forsite Diagnostics Ltd, York, U.K.) and UNISCAN™ Immunoassay Rapid Test Reader (Unison Biotech, Taiwan) can read colorimetric and fluorescent assays. Currently, LRE and Wiagen readers lead the market in fidelity and sensitivity. These semi-quantitative lateral-flow assay readers offer adequate sensitivity, but are high cost ($1–$2 k). All the readout devices outlined are test-strip specific, in other words not standard reading platforms. The smartphone application described in this thesis has the potential to contribute to the development of universal and connected readout devices.

7.5 The Vision for Holographic Sensors

Sensing mechanisms for point-of-care tests are primarily based on gold colloids with antibody/antigen interactions, molecular-dye-based sensors and electrochemistry [52]. These formats are ubiquitous and universally applicable for the use of simple, qualitative and low-cost point-of-care applications, while also having enough capability to be utilised in sensitive, quantitative and multiplex assays. Therefore, academic efforts should focus on holographic sensing applications that are not currently feasible with the existing sensing platforms and explore areas in great need. These niche areas are reusable, implantable, wearable, wireless and powerless devices, possibly targeting analytes at μ/mM concentrations. The ultimate challenge will be to justify the performance and the cost to attain a potential "killer" application, which was achieved through urine dipstick of molecular-dye-based sensing, pregnancy test of NP-based assays, and blood glucose monitoring of electrochemical sensing. Other commercialisation routes may require the developers to embrace the holographic sensors as an enabling platform, which might be an integral part of another sensing technology. The holographic sensors offer unique attributes since they not only provide the interrogation and reporting transducer, but they also have the analyte-responsive hydrogel, rendering them label-free and reusable with remote sensing capability [53]. The single-pulse laser patterning represents the first step towards producing multiplex hydrogel-based holographic sensors that can display 3D images. The future of holographic sensing is at the interface of photopolymerisation, printing, digital holography and integration with microfluidic devices, implantable chips and contact lenses. It is envisioned that holographic sensors will find applications from in vitro diagnostics to dynamic displays to security devices (Fig. 7.3).

Fig. 7.3 Potential
applications of holographic
sensors range from in vitro
diagnostics to optical devices

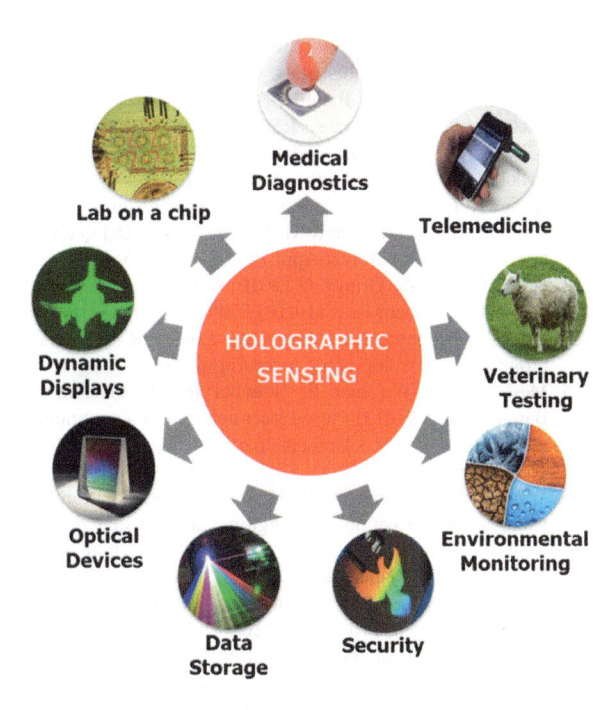

References

1. Yetisen AK, Butt H, da Cruz Vasconcellos F, Montelongo Y, Davidson CAB, Blyth J, Chan L, Carmody JB, Vignolini S, Steiner U, Baumberg JJ, Wilkinson TD, Lowe CR (2014) Light-directed writing of chemically tunable narrow-band holographic sensors. Adv Opt Mater 2 (3):250–254. doi:10.1002/adom.201300375

2. Tsangarides CP, Yetisen AK, da Cruz Vasconcellos F, Montelongo Y, Qasim MM, Wilkinson TD, Lowe CR, Butt H (2014) Computational modelling and characterisation of nanoparticle-based tuneable photonic crystal sensors. RSC Adv 4(21):10454–10461. doi:10.1039/C3RA47984F

3. Yetisen AK, Qasim MM, Nosheen S, Wilkinson TD, Lowe CR (2014) Pulsed laser writing of holographic nanosensors. J Mater Chem C 2(18):3569–3576. doi:10.1039/C3tc32507e

4. Yetisen AK, Montelongo Y, da Cruz Vasconcellos F, Martinez-Hurtado JL, Neupane S, Butt H, Qasim MM, Blyth J, Burling K, Carmody JB, Evans M, Wilkinson TD, Kubota LT, Monteiro MJ, Lowe CR (2014) Reusable, robust, and accurate laser-generated photonic nanosensor. Nano Lett 14(6):3587–3593. doi:10.1021/nl5012504

5. Marshall AJ, Blyth J, Davidson CA, Lowe CR (2003) pH-sensitive holographic sensors. Anal Chem 75(17):4423–4431. doi:10.1021/ac020730k

6. Vasconcellos FD, Yetisen AK, Montelongo Y, Butt H, Grigore A, Davidson CAB, Blyth J, Monteiro MJ, Wilkinson TD, Lowe CR (2014) Printable surface holograms via laser ablation. ACS Photonics 1(6):489–495. doi:10.1021/Ph400149m

7. Blyth J (1985) Security display hologram to foil counterfeiting. In: Huff L (ed) Applications of holography. SPIE—The International Society for Optical Engineering, Los Angeles, CA, pp 18–23

8. Sang L, Zhao Y, Burda C (2014) TiO_2 nanoparticles as functional building blocks. Chem Rev 114(19):9283–9318. doi:10.1021/cr400629p

9. Cargnello M, Gordon TR, Murray CB (2014) Solution-phase synthesis of titanium dioxide nanoparticles and nanocrystals. Chem Rev 114(19):9319–9345. doi:10.1021/cr500170p

10. Deng S, Yetisen AK, Jiang K, Butt H (2014) Computational modelling of a graphene fresnel lens on different substrates. RSC Adv 4(57):30050–30058. doi:10.1039/C4ra03991b

11. Allen MJ, Tung VC, Kaner RB (2010) Honeycomb carbon: a review of graphene. Chem Rev 110(1):132–145. doi:10.1021/cr900070d

12. Kong X-T, Butt H, Yetisen AK, Kangwanwatana C, Montelongo Y, Deng S, Cruz Vasconcellos FD, Qasim MM, Wilkinson TD, Dai Q (2014) Enhanced reflection from inverse tapered nanocone arrays. Appl Phys Lett 105(5):053108. doi:10.1063/1.4892580

13. Hu L, Hecht DS, Gruner G (2010) Carbon nanotube thin films: fabrication, properties, and applications. Chem Rev 110(10):5790–5844. doi:10.1021/cr9002962

14. Blyth J, Millington RB, Mayes AG, Lowe CR (1999) A diffusion method for making silver bromide based holographic recording material. Imaging Sci J 47(2):87–91

15. Messina GC, Wagener P, Streubel R, De Giacomo A, Santagata A, Compagnini G, Barcikowski S (2013) Pulsed laser ablation of a continuously-fed wire in liquid flow for high-yield production of silver nanoparticles. Phys Chem Chem Phys 15(9):3093–3098. doi:10.1039/C2cp42626a

16. Wagener P, Ibrahimkutty S, Menzel A, Plech A, Barcikowski S (2013) Dynamics of silver nanoparticle formation and agglomeration inside the cavitation bubble after pulsed laser ablation in liquid. Phys Chem Chem Phys 15(9):3068–3074. doi:10.1039/C2cp42592k

17. Naydenova I, Jallapuram R, Toal V, Martin S (2008) A visual indication of environmental humidity using a color changing hologram recorded in a self-developing photopolymer. Appl Phys Lett 92(3):031109. doi:10.1063/1.2837454

18. Naydenova I, Jallapuram R, Toal V, Martin S (2009) Characterisation of the humidity and temperature responses of a reflection hologram recorded in acrylamide-based photopolymer. Sens Actuators B 139(1):35–38. doi:10.1016/j.snb.2008.08.020

19. Mikulchyk T, Martin S, Naydenova I (2013) Humidity and temperature effect on properties of transmission gratings recorded in PVA/AA-based photopolymer layers. J Opt 15 (10). doi:10.1088/2040-8978/15/10/105301

20. Mikulchyk T, Martin S, Naydenova I (2014) Investigation of the sensitivity to humidity of an acrylamide-based photopolymer containing N-phenylglycine as a photoinitiator. Opt Mater 37:810–815. doi:10.1016/j.optmat.2014.09.012

21. Farandos NM, Yetisen AK, Monteiro MJ, Lowe CR, Yun SH (2014) Contact lens sensors in ocular diagnostics. Adv Healthc Mater. doi:10.1002/adhm.201400504

22. Blyth J, Lowe CR, Davidson CAB, Kabilan S, Dobson CA (2005) Holographic sensor. 2005012884 A1

23. Schnars U, Jueptner W (2005) Digital holography: digital hologram recording, numerical reconstruction, and related techniques. Springer, Heidelberg

24. Picart P, Li J (2013) Digital holography. ISTE/Wiley, New York

25. Makos MA, Omiatek DM, Ewing AG, Heien ML (2010) Development and characterization of a voltammetric carbon-fiber microelectrode pH sensor. Langmuir 26(12):10386–10391. doi:10.1021/la100134r

26. Huang BR, Lin TC (2011) Leaf-like carbon nanotube/nickel composite membrane extended-gate field-effect transistors as pH sensor. Appl Phys Lett 99(2):023108. doi:10.1063/1.3610554

27. Modi S, Swetha MG, Goswami D, Gupta GD, Mayor S, Krishnan Y (2009) A DNA nanomachine that maps spatial and temporal pH changes inside living cells. Nat Nanotechnol 4(5):325–330. doi:10.1038/Nnano.2009.83

28. Tantama M, Hung YP, Yellen G (2011) Imaging intracellular pH in live cells with a genetically encoded red fluorescent protein sensor. J Am Chem Soc 133(26):10034–10037. doi:10.1021/Ja202902d

29. Grover A, Schmidt BF, Salter RD, Watkins SC, Waggoner AS, Bruchez MP (2012) Genetically encoded pH sensor for tracking surface proteins through endocytosis. Angew Chem Int Edit 51(20):4838–4842. doi:10.1002/anie.201108107

30. Novell M, Parrilla M, Crespo GA, Rius FX, Andrade FJ (2012) Paper-based ion-selective potentiometric sensors. Anal Chem 84(11):4695–4702. doi:10.1021/ac202979j
31. Asher SA, Sharma AC, Goponenko AV, Ward MM (2003) Photonic crystal aqueous metal cation sensing materials. Anal Chem 75(7):1676–1683. doi:10.1021/ac026328n
32. Baca JT, Finegold DN, Asher SA (2008) Progress in developing polymerized crystalline colloidal array sensors for point-of-care detection of myocardial ischemia. Analyst 133 (3):385–390. doi:10.1039/B712482a
33. Jiang HL, Zhu YH, Chen C, Shen JH, Bao H, Peng LM, Yang XL, Li CZ (2012) Photonic crystal pH and metal cation sensors based on poly(vinyl alcohol) hydrogel. New J Chem 36 (4):1051–1056. doi:10.1039/C2nj20989f
34. Yetisen AK, Montelongo Y, Qasim MM, Butt H, Wilkinson TD, Monteiro MJ, Lowe CR, Yun SH (2014) Nanocrystal bragg grating sensor for colorimetric detection of metal ions (under review)
35. Zhang J-T, Wang L, Luo J, Tikhonov A, Kornienko N, Asher SA (2011) 2-D array photonic crystal sensing motif. J Am Chem Soc 133(24):9152–9155. doi:10.1021/ja201015c
36. Ben-Moshe M, Alexeev VL, Asher SA (2006) Fast responsive crystalline colloidal array photonic crystal glucose sensors. Anal Chem 78(14):5149–5157. doi:10.1021/ac060643i
37. Yang X, Lee MC, Sartain F, Pan X, Lowe CR (2006) Designed boronate ligands for glucose-selective holographic sensors. Chemistry 12(33):8491–8497. doi:10.1002/chem.200600442
38. Kabilan S, Marshall AJ, Sartain FK, Lee MC, Hussain A, Yang XP, Blyth J, Karangu N, James K, Zeng J, Smith D, Domschke A, Lowe CR (2005) Holographic glucose sensors. Biosens Bioelectron 20(8):1602–1610. doi:10.1016/j.bios.2004.07.005
39. Kabilan S, Blyth J, Lee MC, Marshall AJ, Hussain A, Yang XP, Lowe CR (2004) Glucose-sensitive holographic sensors. J Mol Recognit 17(3):162–166. doi:10.1002/jmr.663
40. Horgan AM, Marshall AJ, Kew SJ, Dean KES, Creasey CD, Kabilan S (2006) Crosslinking of phenylboronic acid receptors as a means of glucose selective holographic detection. Biosens Bioelectron 21(9):1838–1845. doi:10.1016/j.bios.2005.11.028
41. Yang XP, Pan XH, Blyth J, Lowe CR (2008) Towards the real-time monitoring of glucose in tear fluid: holographic glucose sensors with reduced interference from lactate and pH. Biosens Bioelectron 23(6):899–905. doi:10.1016/j.bios.2007.09.016
42. Marshall AJ, Young DS, Kabilan S, Hussain A, Blyth J, Lowe CR (2004) Holographic sensors for the determination of ionic strength. Anal Chim Acta 527(1):13–20. doi:10.1016/j.aca.2004. 08.029
43. Yetisen AK, Akram MS, Lowe CR (2013) Paper-based microfluidic point-of-care diagnostic devices. Lab Chip 13(12):2210–2251. doi:10.1039/c3lc50169h
44. Volpatti LR, Yetisen AK (2014) Commercialization of microfluidic devices. Trends Biotechnol 32(7):347–350. doi:10.1016/j.tibtech.2014.04.010
45. Yetisen AK, Volpatti LR (2014) Patent protection and licensing in microfluidics. Lab Chip 14 (13):2217–2225. doi:10.1039/c4lc00399c
46. Akram MS, Daly R, Vasconcellos FC, Yetisen AK, Hutchings I, Hall EAH (2015) Applications of paper-based diagnostics. In: Castillo-Leon J, Svendsen WE (eds) Lab-on-a-chip devices and micro-total analysis systems. Springer, Dordrecht
47. Yetisen AK, Jiang L, Cooper JR, Qin Y, Palanivelu R, Zohar Y (2011) A microsystem-based assay for studying pollen tube guidance in plant reproduction. J Micromech Microeng 21 (5):054018. doi:10.1088/0960-1317/21/5/054018
48. Yetisen AK, Martinez-Hurtado JL, Garcia-Melendrez A, Vasconcellos FC, Lowe CR (2014) A smartphone algorithm with inter-phone repeatability for the analysis of colorimetric tests. Sens Actuators B 196:156–160. doi:10.1016/j.snb.2014.01.077
49. Yetisen AK, Martinez-Hurtado JL, da Cruz Vasconcellos F, Simsekler MC, Akram MS, Lowe CR (2014) The regulation of mobile medical applications. Lab Chip 14(5):833–840. doi:10. 1039/c3lc51235e
50. Shen L, Hagen JA, Papautsky I (2012) Point-of-care colorimetric detection with a smartphone. Lab Chip 12(21):4240–4243. doi:10.1039/c2lc40741h

51. Hong JI, Chang B-Y (2014) Development of the smartphone-based colorimetry for multi-analyte sensing arrays. Lab Chip 14(10):1725–1732. doi:10.1039/C3LC51451J
52. Chin CD, Linder V, Sia SK (2012) Commercialization of microfluidic point-of-care diagnostic devices. Lab Chip 12(12):2118–2134. doi:10.1039/c2lc21204h
53. Yetisen AK, Naydenova I, Vasconcellos FC, Blyth J, Lowe CR (2014) Holographic sensors: three-dimensional analyte-sensitive nanostructures and their applications. Chem Rev 114 (20):10654–10696. doi:10.1021/cr500116a

CPSIA information can be obtained at www.ICGtesting.com
Printed in the USA
LVOW05*1803231214

420157LV00001B/78/P